RETURN TO CREATION

By
E. Clifton Sosebee

<u>Dedication</u>

Writing science-fiction will always be one of my greatest loves, because it is the most versatile genre that can be written about due to the endless, uncountable mysteries that can be formulated by its vast unfamiliar unknowns that we have only begun to barely understand.

However, I like to keep my Sci-Fi stories at least as close as I can within most of the believable boundaries of Mother Nature's unique engineering mechanics as possible. At least as close as I can to the universal laws of physics that control our universe, which to me keeps the story more realistic, that shapes the story into being much more believable to understand.

A scientist I've come to know, who is not only a good friend that helped me along with the vast scientific knowledge of universal astrophysics, kept me within most of the boundaries of our own beliefs of the true common laws of celestial mechanics and the physics that only God Himself governs their unpredictable actions.

And it was not just the abundant knowledge of how the immense engineering of these celestial mechanics of how our grand cosmos performs its universal functions; it was mostly the scientific supportive viewpoints to my ideas; some of which I had not the slightest clue of what might happen if certain unknown events were to transpire.

I would like to take this moment to thank this scientist so much for taking your valuable time to guide me along with the scientific details of this novel that I had so much fun writing knowing you were standing on the sidelines during certain scenarios that an unknown journey through unfamiliar space could bring about.

It was a special privilege to write this book knowing I had you helping me along the way…But it was a much greater honor that someone of your vast colorful character would give me the proud, respectful admiration that you would allow me to be your friend.

Dr. Michelle Thaller
****

Author's Note

When I was a little boy growing up in Ohio, one of my greatest memories was going outside just after sundown with my father, and we would sit on the side of our house just overlooking Keller's Hill to watch the stars come out.

He would then guide me through all the wonderful stars and planets, and as the love swiftly grew inside that little boy for the magnificent amazement he was looking at, I would hammer my Dad with increasingly more questions the more he taught me as time went on.

To this very day that same little boy inside me still continues to ask about the marvels of our universe, and because of my wonderful Dad, one of my greatest loves will always be the unbelievable mysteries of physics.

Even as a child, one of those questions that will always burn inside me cannot be answered during this time of our present technology, nor will it be for uncounted centuries yet to come in the far distant future.

The question is what's that behind the moon, planets and stars, and why is it all black? I've come to realize since that time a much better way to ask that unique question is what lies beyond what we cannot see.

No doubt the day will come when our civilization will discover everything that engineers and operates the universe that we live in, for our scientific technology has advanced so geometrically that scientists have discovered more in the last twenty years than in the last hundred, and more in the last hundred than we did in the last ten thousand.

Once that day comes when scientists know everything there is to know about our universe, then their curiosities will turn toward answering the next question many scientists have asked since their own childhoods: what lies beyond the darkness of space that scientists are now beginning to believe is not infinite; that some kind of boundary does exist somewhere in the immeasurable, remote distance of deep black space.

And who knows, perhaps there will come a day, if we don't stumble upon it somehow during the journey seeking as to what lies beyond, that scientists will find the answer to the greatest question since conscious thought first made its initial introduction of awareness into the universe's 13.7 billion year existence:

Why?

Why are we here, and for what purpose?

In my opinion, the answer to that is so extraordinarily remarkable, it is something at this stage we cannot even think about or try to possibly understand, or much less know where to begin an attempt to understand or comprehend.

Until such time, we'll have to settle with what our incredible scientists have learned; that for some reason we exist in a universe that all we really know for certain is that we are here.

But at this so young and early stage of discovery regarding our present progress of scientific evolution, to prove just how far we have to go, we really do not have the slightest clue where here even is.

"All I ask is a tall ship,

And a star to steer by."

<u>*John Masefield*</u>

CHAPTER 1

CORPORATE HIJACK

Watching the captivating tempests churning their way across the face of the magnificent gas giant's beautiful blue surface always intrigued him, knowing the mighty currents of Neptune's unyielding atmospheric airstreams were the most powerful prevailing winds above any other planet in the solar system.

His hypnotic trance was softly interrupted by the voice of the shuttle pilot sitting just next to him, "That's one of the most incredible views I could never get tired of watching, Dr. Stanley. It's so tranquil to gaze across that magnificent, colorful surface of the majestic rich blue color of Neptune's serene looking surface."

"Indeed, Holly," he calmly replied to the pilot, "That serene looking surface is unfortunately masqueraded with a tremendous deception of titanic, bigger than earth size cyclones gradually creeping their way across the planet's atmosphere."

The little transport shuttle leisurely began to ease to the right, preparing to swing its way around Neptune to the planet's opposite side, where their destination was orbiting one of the massive planet's outer most moons.

Once they had turned enough to clear the planet's long curving crescent, they gradually started to shift back left, now swerving around to Neptune's immense backside.

They soared passed the beautiful breathtaking feature of the inner moon Triton. Holly readjusted the shuttle's heading, now heading out away from the magnificent planet toward the long eccentric orbits of the outer moons, and their objective.

"So how many trips is this for you out to the Hubble station, Doctor?"

"I lost count," he answered reluctantly, "Not that I mind coming. I mean who would? But never at such a quick notice to drop whatever I'm doing and get out here as quickly as possible, without so much as a reason why. It's really strange."

"Something's up," Holly said, "I mean we have been on a round the clock schedule getting every engineering technician available out here for almost two months now, especially the top scientists like you."

He turned to her and asked with a tone of surprise, "Really? Just how many of us?"

"All five of you, Doctor," she replied, "But you're the last one. I can't wait to get there, so I can finally get a good night's sleep."

He turned his confused gaze back forward again, and said, "That's even stranger. I can't wait to get there either to find out what the big secret's about. That's just as odd as when they first notified me to get out here pronto. Every time I asked what's the big rush, the only response I got is that I'll be fully briefed once I've arrived. They've never held back any classified info to me."

"Guess you'll know soon enough, Dr. Stanley. Look," she said as she raised her arm and pointed out of the shuttle's front viewport.

The moon Nereid's circular shape began slowly creeping into their sight. Moments later their entire view began to reveal one of mankind's present technological achievements of the twenty-fifth century, as the shuttle closed in on the incredible Hubble VII space station, majestically moving in its high orbit around Neptune's outer moon.

Holly eased the shuttle's velocity down while adjusting the little spacecraft's motion to match that of the enormous station's circular landing podium, gradually maneuvering the shuttle downward to rest dead center of the gleaming silver platform.

"That's one view I'll never get tired of seeing," Dr. John Stanley said to Holly in low soft tone of admiration, "Our new eyes to the known universe. Absolutely magnificent."

A small jolt shook throughout the shuttle's cabin, indicating the little craft was now being lowered down into the shuttle bay disembarking area. Yet John's gaze of captivating amazement never shifted away from the spectacular view of the new Hubble VII telescope's tremendous duel cylinders, both seven hundred feet in diameter, protruding outward from the vertical opening in the enormous domed structure where the JPL staff operated it from.

"Biggest set of binoculars I've ever seen," Holly remarked as the elevator plank lowered the shuttle down below the landing platform. Another similar jolt trembled within the cabin, signifying they had arrived in the station's main shuttle bay.

They walked to the rear of the shuttle just as the rear exit door divided apart. Stopping for a brief moment waiting for the rear exit's steps to swing down to the main hanger deck's floor, John turned and said, "Thank you Holly. It was nice to see you again."

Holly gestured her hand out to shake his, and replied, "The pleasure's always mine, Dr. Stanley. I'm just glad it was me that had the honor of bringing out NASA's number one scientist."

"You get some rest now girl," and stepped sideways to motion ladies first, as any courteous gentleman would.

Several steps before he reached the shuttle's main deck, he recognized the senior ranking station chief executive, and one of his long time career closest friends, Barry Harper. To his surprise, as Barry stepped up to greet him, there were two of the station's security personnel standing alongside and keeping exact pace with him.

"I knew I couldn't kill bad grass," Barry shouted, and continued while greeting John with a firm handshake, "Because you always grow back! How's it going, John? Keeping that staff of yours in line at NASA?"

"Only when I know where they are. When I was kidnapped to come here, I wasn't even allowed to contact NASA back home to let any of them know. Are any of them here?"

"All four of them," Barry answered, "We kidnapped them too. C'mon."

"What? Are we in some kind of trouble?"

Barry smiled back, and replied, "Depends on your point of view. Let's go. They should be in the main conference room by now. We were all waiting for you."

Walking through the hanger deck with the security officers on each side of them, John was almost in a state of shock. He mumbled to Barry, "What's with the special escorts? Station security lockdown?"

"Special project lockdown, old friend," Barry softly said back, "Yours in fact."

"What? Just what kind of project?"

Barry leaned over to John as they walked, and softly said, "Out of everything I've ever wished for, this one has to be my greatest wish of all."

"Sounds big," John said, "So what's the big wish?"

Smiling from ear to ear, Barry looked at his old companion as they walked, and said almost in a tone of whisper, "The look on your face when you hear the whole picture."

CHAPTER 2

CAUSE AND EFFECT

When they arrived at the main conference room door, Barry then turned to the two security men and said, "Wait out here."

Barry turned and opened the conference room door, then motioned to John with a little tilt of his head toward the room, smiling even more now.

John stepped in and came to a quick halt. There they were, his four ingenious teammates, all in their atypical eccentric little worlds that made them so extraordinarily special.

A man that looked to be in his early thirties sitting to the far left side had his head tilted straight back facing the ceiling, swaying his head slowly side to side while rapidly twirling an ink pen between his fingers on each hand.

To his right sat two women, one very young woman mildly arguing with another sitting to her right that looked to be in her late twenties, always challenging each other with hypothetical theories usually about astrophysics, or anything else both could put out the effort to take on the other.

To the far right side sat John's most special colleague, a very young man with two small hand computers in each hand with his thumbs dancing across the touch screen monitors so fast it looked like a cloudy shapeless blur.

When Barry stepped through the door, none of them even noticed their supervisor or the station's chief administrator had come in. John glanced back to Barry with a proud, silent smile. Barry returned his glance with a grim look of impatience, and then stepped back and slammed the door shut.

A spinning ink pen went flying across the room as it slipped out of the startled man who was sitting on the left side's right hand, but was able to grasp the one in his left hand in time. He slowly stood up along with the others, and was the first to speak by shouting, "Boss man! It's about time you showed up!"

The four of them rose and walked around from the rear table up to John, while John said in a loud tone to Barry, "I guess it's safe to assume you've already met my colorful crew here."

"Not formally, John. When we brought them in, they were even more annoyed than you were."

John snapped back, "Were? I still am, until we find out what this is all about."

He turned back to greet them, and said as he shook hands with the first one that walked up, "Hello, Jerry. I see you're still at your circus act with those pens. Doing okay?"

"Pretty much, boss. I'll be doing much better when they finally tell us why we were all pirated here like fugitives."

John turned back to Barry, and said, "This is Dr. Jerry Danielson, Barry, our analysis technician and ace survey pilot when necessary. Jerry, Barry Harper, the Hubble VII station's chief administrator."

They shook hands while John continued, "And this is Dr. Miranda Young, our special projects technician, and this young beauty is Dr. Barbara McKenzie, our deep space astrophysicist expert, and number two computer wizard."

The two ladies just nodded toward Barry along with a short wave.

John then turned to the last of the four, and said, "And this young man is our other astrophysicist expert, but also our number one super genesis computer specialist, Dr. Steven Meyers."

Steven stepped up to Barry, shook his hand and very politely said, "Hello, Mr. Harper. It's an honor to meet you sir."

"Thank you Steven," Barry returned his nice gesture with, and then turned to the rest of them, and said, "Thank you all. Please, everyone have a seat. I'm sure all of you all are eager to know what the big mystery is all about."

They all returned back to the rear table sitting were they originally were, except for John that seated himself just to the right of the two ladies.

There was a small desk in the front of the big room that Barry scooted himself onto the top of the middle, grasped both hands to the front of its front edge, and leaned forward smiling at all of them with some kind of confusing sense of pride.

His smile grew even larger with each passing moment, almost to the point where he couldn't get it out whatever it was he had to tell them, just staring at NASA's top five scientists now grinning even more, but still in complete silence.

Without turning his head, John's eyes shifted quickly from side to side to see if any of his teammates were beginning to get annoyed as he was. The other four were

14

frozen stiff, but shifting their eyes back and forth from John to Barry wondering when this meeting was ever going to start.

A few moments later, John sighed loudly enough for Barry to hear, leaned back in his seat, and then firmly said, "We're not getting any younger Barry. Anytime."

The station manager then finally spoke, and said, "Sorry, John, it's just that I'm seeing what five people's faces look like that are in the last few seconds of the last chapter of their past lives, and want to compare my image of what your five faces will look like when the next chapter starts. Oh, God, my good friends, how I so wish I could be one of your five faces in a few moments from now, and all the moments that will follow for the rest of your lives!"

The four NASA subordinates swiftly jerked their serious glares of concern directly to their boss. Miranda Young leaned her head down far enough so she could look around John over to Jerry, and shouted, "You see, damn it Jerry! I told you she was lying when she said she was over eighteen! But no, you go and let those ballistic hormones of yours get us all…"

John instantly held his hand up to her to stop, while another screaming voice on his opposite side retaliated, "What in the hell are you talking about Miranda? I didn't do anything but shoo her away when she tried to come on to me!"

"Stop it! Both of you," John roared, and then swung his head to the right and left at both of them, and yelled, "Now! I don't want to hear another word!"

He then turned toward the front of the room rising to his feet, and shouted once again, "Except from you! This mystery, hush-hush maneuver you and the rest of the military are pulling on the five of us Barry has gone far enough!"

Barry had already slid off the desk and was walking up to John, still smiling from ear to ear. He stopped directly in front of John, and proudly said, "In that case, my dear old friend, please allow the first word you're going to allow me to say, is congratulations."

Standing face to face with his old companion, John hesitated for a couple of seconds, and then asked in a smaller tone of anger, "For what? What are you talking about?"

"Cause and effect, John," Barry answered.

Barry then took a couple of steps back, and said to everyone, "Congratulations to all of you. You've done it, and John, what you and your magnificent team have accomplished, saying this has gone far enough in the scope of what I'm about to tell you, is on the right track but quite inaccurate. You have no idea just how, "far," this is going to take the five of you."

CHAPTER 3

THE TRIUMPHS OF TECHNOLOGY

"Fair enough. Let's have it."

"You and your team's formula, John," he said while scooting himself up on the long table just on the other side of where John and his team sat, "It works. We've been testing it for three months now. Our engineers have sent five probes out now using the incredible force of power your team submitted to General Vaughan five months back. How you guys found a way to capture and detonate tachyons has…"

Little Steven Meyers stepped right in, "You mean harness, stabilize and implode tachyon emissions?"

John followed with his tone of anger rising up again, "Wait a minute! Just wait one damn minute! We presented that to the Department of Defense five months ago, after working for almost two years on Steven and Barbara's theory. After a week I didn't get one call from anyone at the Pentagon, so I contacted him General Vaughan about it, and he acted like he didn't know what I was talking about!

"So I asked Steven and Barbara to resubmit it again, and when they went to pull it back up, we find it had been completely wiped clean out of our system! Steven said only one organization anywhere had the capability of doing that! The gifted scientists at the Pentagon in our own country's

Special Project Commission, under the command of General Vaughan!"

Barry kept waving his hands in front of John while shouting back a quick, "Please, hear me out."

But John continued to pour it on, "I should have known! I can't believe this! Those insecure, military morons found...no, stole, a new destructive power to play with, and you keep it from us because, wait, let me guess, a possible threat to national security? You let me tell you something, my so trusted and dear 'friend'! There's no way..."

Laying all the way back now on the long table with both his hands over his ears, knowing the whole time this was coming and was well merited, he hoped that soon John would begin to wear down to a point that he could finally tell his old friend what this whole picture was about, but he had promised himself that in no way would he reveal the great news without first telling him the truth about the government wrongly withholding his friend and his team's news about this incredible new discovery.

To reveal the extraordinary news first before telling him the bad he was certain was the wrong way, and that by doing it the right way it would somehow pay off.

Doing things the right way instead of the easy had always paid off for Barry Harper, and with the scope of such magnitude as in this case, that which controls the fate of all that is good finally fell right in his lap.

Suddenly he heard a distant shout over John's muffled screaming followed by instant silence.

Leaning quickly up off the table, everyone's attention was toward the front of the room, where the two guards had come in and were halfway across the room walking quickly toward him.

The ranking security officer was shouting, "Sir! Is there a problem we can assist you and Dr. Stanley with?"

"The only problem here, Lieutenant," Barry swiftly sliced into his oncoming question with, "Is that you two are away from your posts, which can be resolved by you resuming them."

"Yes sir," the ranking officer replied. He then motioned to the subordinate officer as they stridently vacated the room.

"Okay, great," Barry said with a long sigh of frustration. He then asked, "Anybody up to some lunch before we get back to biz?"

The four of them shouted everything from cheeseburgers to roast beef sandwiches, but quickly John overruled them, and said absolutely not until this meeting was over, and he had his answers.

CHAPTER 4

FORTUNATE IMPATIENCE

The stations' CEO then turned back to face the blank stares of silence from all five sists, and calmly asked, "May I now speak? Please?"

Not so much as a wink was his response from the five cold glares.

"Thank you," Barry politely said, and then began to leisurely pace in front of the conference table again.

"John, your reaction was expected to be far worst, because truthfully mine would have been totally ballistic. Now would you please allow me to finish what I have to say, because like I've been insinuating, what's in store for you and your team is too incredible to believe when I tell you. Trust me, the only way you're going to fully accept that what I'm about to tell you is really true, you will all have to still see it to believe it."

Still in a cold, smooth response of anger, John said, "The anticipation of finally getting out of you the bottom line explanation for this entire charade is the only thing that's keeping me from strangling you to death right now. Why can't you just come out with everything at once?"

"Alright," Barry pleaded. "From the top. General Vaughan received your theoretical discovery, and was so impressed he had his people drop everything and gave the testing project top priority. They managed to capture and contain three tachyon particles, but it took them two days to do it. General Vaughan had run out of patience, and…"

Steven stood quickly up, and said, "Sir, if they had shielded the containment field with copper based lining using a minimum of ten centimeters they could have captured hundreds of tachyons every second. Sir."

"After two days," Barbara added, "They would have had billions of tachyons incased in the containment field. Steven and I sent a fully illustrated schematic diagram. Why didn't they follow those instructions?"

Barry only shrugged his shoulders, and answered, "Regretfully my dear young lady, when a new type of energy to any establishment is initially discovered, the most prominent factor that presides over everything is the impatience of the politics that govern it. Once General Vaughan saw the degree of immeasurable enormity it generated, he had his special project people practically running into each other to assemble all the factors together for the initial test.

"I'll give you a perfect example. What happens when a young boy gets a really cool toy for Christmas? When he rips the box apart to feel the joy of seeing the new toy, the first thing he throws aside is the instruction manual. Except in this case, had they gone by you and Dr. Meyers's guidebook, the results would have been so extremely devastating, chances are we would not be having this conversation."

The five NASA scientists all looked at each other now with utter confusion.

John's eyes then widened with a sharp glare of recognition, and softly said to his colleagues, "He's right. I understand perfectly what he's saying people. Do the math. Go on Barry."

"Three months later a survey vessel placed the device they assembled for the test that contained the three imprisoned tachyons on the surface of Uranus's moon Oberon. The ship then swung around the opposite side of Uranus, engaged the device's containment field to release the three tachyons into the device's zero vacuum inner core.

"What they monitored then is actually too inconceivable. When the three tachyons instantly collided together the monitor and every reading covering the test went dark for maybe two seconds. But the velocity they were traveling at just before they all collapsed together, when all contact with the device readings went dark, is too much for my realistic side to ever accept. Seventeen million, four hundred fifty one thousand, seven hundred seventy one miles per second….."

"That's seventeen million," Steven politely leaped in, "Four hundred fifty one thousand, seven hundred seventy one point zero zero zero seven three nine three four miles per second….. Um, sir."

Smiling with his great sense of pride for accuracy, he noticed everyone, even Barry Harper giving him a cold glare of annoyance.

The always happy–go–lucky young scientist just rolled his eyes down while mumbling, "Sorry. Just thought, well, you know….."

"So the ship," Barry resumed, "Rounded its way back to the opposite side of the planet to check the device for some kind of malfunction, but it was gone."

John asked, "That's strange. What do think happened to the device?"

Followed by a sarcastic little laugh, Barry stepped up to the other side of the table where John sat, placed his hands down on the table and leaned toward John, and replied, "I'm not talking about the device old buddy. I'm talking about the entire moon Oberon. It no longer exists."

CHAPTER 5

BLIND ACCEPTANCE

In the few succeeding instants that followed seemed like one of those rare moments in time that you could live in forever. The five NASA scientists sat with the silence of sheer awestruck triumph.

Steven wanted to speak, but the glory of that magnificent moment had him completely paralyzed as his voice remained motionless, so he gradually raised his hand waiting for the Hubble VII CEO to concede him.

With his elbow stretched out across the table with his curled fist supporting his resting cheek, it was the voice of Jerry Danielson that first breached that soft air of serene concern, "Great. Work all my life to get where I am now, only to become a victim of poetic justice. Now I'm going to be locked up for the rest of my life for conspiracy to commit moon murder."

"Shut up, Jerry," echoed swiftly throughout the large room from surprisingly little Barbara McKenzie, who then rose from her chair, walked over and scooted herself sitting sideways across the long table just next to Steven.

She then looked down into Steven's glazing watered eyes that matched hers, with the exception of one small tear making its way down one of her cheery cheeks.

Barbara then leaned over just enough to encircle her arms around him, and they both embraced each other so hard from the utter joy of success, Steven didn't care that he was losing his breath as the tears began to roll down his face when Barbara screamed to the top of her voice, "Yes! Oh my God Steven! Yes! We did it! Oh my God!"

That broke John Stanley's frozen glare away from the smiling face of his old friend, now shifting his look of shock over to Miranda then back to the other side at Jerry.

He then leisurely raised his chin back up to Barry Harper, and said with almost a serene whisper, "I can't believe it."

Another soft voice of complete astonishment from Miranda, "With the implosion resulting from the collision of only three tachyon emissions?"

Barry continued his glorious smile as he acknowledged her by slowly rolling his head up and down, but still looking at the face of bewildering amazement coming from his long time friend.

Barbara's new voice of authority then sharply struck Barry, "Mr. Harper, when Steven and I put together that discovery proposal, we made it very clear sir, very clear in our cover profile that accompanied this hypothetical proposition, to never, ever under any circumstances let the collected tachyons accumulated in whatever type of instrumental containment field ever designed to process and contain them, allow more than two, and only two at one time to be released into the pressurized chamber of the inner core constructed to bring them together."

"Miss McKenzie," Barry started to say, "You're not listening…"

"No, Mr. Harper," shouted Miranda as she leaped to her feet and began rounding the table she was sitting behind heading toward the front, "It's you that is not listening! I know what Barbara's trying to explain, and…"

John quickly snapped out, "Miranda! Back off…"

"One minute," she rebounded back to her boss pointing her outstretched arm's finger toward him but never taking her piercing eyes off Barry, but taking a split second to jerk her glare his way to add, "Sir!"

She then snapped her head back to Barry, and resumed her attack, "You will let her finish what she's trying to get across to you, because those idiots you people call scientists that took shortcuts to hurry General Vaughan's new toy along didn't have enough common sense to follow Barbara's and Steven's clear and direct warning, what you refer to simply as an, 'instruction manual,' could have resulted into something so catastrophic that even Armageddon itself couldn't hold a candle to. Got it?"

"Go ahead, Barbara," she then quickly yelled standing face to face to Barry.

Barbara's tiny physique had already made its way up to Barry's opposite side. John Stanley only eased back in his chair with a growing charm of pride watching his two female staff, who argued constantly until an outsider of any type dared to take one of them on that always triggered the true bond of admiration they really held for each other, as the outsider then faced the powerful wrath of both baring down on him.

"Thanks Mir," Barbara said, and then looked up to swollen eyes of fear of Barry, "Now listen to me very carefully, and do not interrupt me until I have finished explaining the entire scope of what those pathetic buttheads almost did."

Jerry had already taken the seat just to the right of John, and quietly said, "Who needs the power of imploded tachyon emissions when you got those two bearing down on you."

John jerked his hand up to Jerry to be silent along with a small growing grin.

Barbara then began, "It would seem they did perform the proper procedure by shielding the containment field with the copper based lining.

"However, it sounds as if they did not use the minimum required amount of ten centimeters thick in construction of the device's containment chamber that would have assembled hundreds of tachyons by each passing second.

"It would seem the containment chamber they designed was only protected I would estimate by two, perhaps three centimeters in thickness at the most. Thus, they were able to capture only three tachyons.

"Steven and I explained in our instruction procedures that only two, and only two tachyons must be fed from the containment chamber into the device's inner core for the implosion method to sustain a controllable yield of kinetic energy.

"This procedure of shielding the containment chamber with only two or three centimeters of copper based lining apparently General Vaughan's team simply ignored. Thank God!"

Barbara then stepped closer to Barry as her voice then began to slowly rise to a deep roar, "Had they pulled that fatal stunt on Oberon by constructing the containment chamber with the proper minimum thickness of ten centimeters, and then released a few billion tachyons at once into the devices' inner core instead of only two at a time, when the high energy protons were then released into the inner core that encircles the gathered tachyons, when released to the vacuum of space and the high energy protons crushes the tachyons under the vast amount of artificial gravitons we imposed to the surrounding protons and triggers the implosion process to that many tachyons at once, the result of that much inconceivable release of kinetic energy, there now would be nothing left but a radioactive cloud of sparkling subatomic particles over a five hundred light year radius of where the Solar System used to be!"

Barry had settled into a state of absolute blind acceptance, slowly moving his head up and down standing directly in front of John's amusing smile he tried so hard not to show. The only part of his body that moved was his eyes, shifting them to each side at the two women wondering if Barbara had finished her verbal assault.

Heaving a long sigh, Barry started to say something, but Barbara's guardian Miranda swiftly yanked her hand in front of him to not yet speak, and asked, "Are you finished yet Barb?"

"No. Just recharging. Now the math. Steven and I submitted the initial proposal to Dr. Stanley five months ago, which he approved and forwarded to General Vaughan's team at the Pentagon."

She then turned toward the long conference table's end where he was sitting while shouting, "Right, Steven? Steven? Where did, oh."

That's when she noticed he had moved to sit right next to John, with Jerry still sitting on John's opposite side, but so close to his superior it was almost as if he was snuggled up to him along with a look of immortal terror directly at

Barbara, as he shouted in instant agreement, "Right! She's right!"

Miranda had to turn away for a moment to keep from bursting out laughing, while Jerry not noticing that Steven had switched seats to a much safer location. He leaned back behind his boss and yelled out with a controlled amusing tenor, "Hey, Steve! Welcome to the party my man! Things getting a little intense for you sitting over there at the end?"

His teasing came to a quick halt when John turned to him with a glare of certain death.

He then turned back toward the front where Barbara and Miranda had Barry flanked from each side, and said, "Ladies, let's calm it down. Come and sit back down."

Barbara shot back, "I just have a couple…"

"Now," John sharply instructed her, and then shifted his look of authority to Barry's other side, "Miranda, please."

They both slowly walked back toward the long table's corner barely hearing Barbara's, "Yes sir." Jerry watched as they turned the corner, and as soon as Barbara passed him, he motioned up to Miranda a gentle pat on the empty chair next to him for her to sit there.

Without looking down, she said as she walked by behind him, "Not a chance in hell."

Barbara sat next to Steven, and as Miranda passed him, she gave him a couple of pats on the back before she sat next to Barbara. She then leaned down enough to look around Barbara toward him, and said, "Hey. You alright kiddo?"

"Yes," was his swift response, "I'm fine. Thank you."

Before she could speak, John gestured to Barbara, "I'll take it from here sweetheart."

He then turned toward his old friend, but before he could speak, Barry headed him off, "I know what you're going to ask John. Since the Oberon incident two months ago, why have we waited this long to tell you? Now comes the easy part, because everything that has transpired in the last two months has all been for the five of you."

Miranda sarcastically snapped out, "Oh, I can't wait to hear this part!"

John again had to intervene, "Miranda! Did I not ask please?"

"It's okay John, because I can't wait to tell you. All of you. Now the fun part begins. Just please excuse me for one quick second."

He then walked to the conference room door, opened it for a brief moment to stick his head out to say something to the guards, and walked back over and stood once again before the frustrating silent glares of the NASA team.

Barry then announced, "There's been a special reception dinner prepared for all you. We'll all be leaving to go there in about ten minutes. This will give me enough time to give you the scoop of what transpired after the Oberon mishap."

"Think I'm ready for a break," John said with a short sigh, "How about you guys?"

After getting an instant full agreement, Barry said, "Great. This is where your payback only starts. Alright. After General Vaughan got word about the Oberon incident, he immediately ordered an emergency full investigation by another government special agency, designed to profile everything starting from the moment he received your first proposal.

"The first thing this new investigation team did was read your, 'fully illustrated schematic diagram,' to the letter. Within three hours General Vaughan had the full story before him as to what the initial project team did not adhere to, that resulted in our third largest gas giant's missing moon.

"The following morning everyone involved in the preliminary first test was put on the street. Twenty seven people are now without jobs. The General was so outraged, he didn't care if there were entry clerks on the first team that had nothing to with the design flaw of the construction of that device, and he wanted them gone as well. I mean he cleaned that entire department out until the floors sparkled with that new refreshing glare to it before he allowed the new, much more professional scientific designers were assembled that he himself handpicked each one."

There was a soft tap on the door. One of security officers then stuck his head in and gave Barry a quick nod.

"Okay, we'll still be a few minutes," he replied, and the door quickly shut.

Barry turned back around and continued, "A little more than a week later, another testing device was ready, built exactly to you and Dr. Meyers' specifications and instructions to the letter. Once completed, another survey craft along with two more, one of which had General Vaughan onboard brought it out to a point somewhere between the orbits of Saturn and Uranus, and deposited it in the vacuum of space. The three ships then backed off five million miles each in a triangular position around the device before detonating it."

He then slipped his hand into the inside of his sports jacket, while at the same time John asked, "Well? What happened?"

While pulling out of his jacket some type of small chrome apparatus with several glowing buttons, he replied, "See for yourself, my friends."

He then pointed the little mechanism toward the front wall of the conference room, snapped one of the buttons and the entire room became completely dark. It was followed by another click within the darkness, and a huge monitor lit up the face of the room's front wall.

A voice could then be heard counting down, saying, "Four, three, two, and one."

Suddenly a blinding white flash leaped out from the monitor, followed by an expanding rolling ring of smooth churning orange that eventually faded away as it spread out into the blackness of space.

Along with every human being that ever lived since the invention of the camera over five centuries ago, there were many times in Barry Harper's life that he wished he had one, but not as much as the moment when he snapped the lights back on.

John, Miranda, Jerry, Barbara and Steven all had glares of catatonic shock, except Barbara's little hands were cupped over her mouth. Steven's fixed stare faced forward still locked on the monitor, with his mouth wide open.

Gradually turning in Barry's direction, John asked, "That, was merely the implosion of only two tachyons?"

Grinning while moving his head up and down, Barry replied, "General Vaughan's strict orders were to abide by Dr. McKenzie's and Dr. Meyers's instructions to the letter.

Anyone that varied from them in any way shape or form would not only permanently lose their position, they would be arrested for insubordination and prosecuted to the fullest extent of the law."

"Holy Mother of God," Miranda softly murmured while rising back out of her trance of shock, then slapping Barbara on the back of her shoulder in a 'way to go' gesture.

"Any Mother of God would be freaked at Barb and Steve's dynamic duo deed like that," Jerry yelled, "Yes indeed! Oh, man, do you guy's rock or what?"

Barbara had leaped out of her chair by then with her arms fully gripped around Steven's neck from his backside, swaying him back and forth screaming at full tilt with triumph into his right ear, but the shy little scientist remained in his fixed catatonic trance with his mouth wide open.

John stood up and shouted with joy, "Then let's start this celebration with that meal you promised us Barry! It's our day for good victory meal!"

"Oh, trust me folks, this dinner they have ready for you is only the tiny tip of the iceberg," Barry shouted as he stepped to his side, and gestured his hand out toward the conference room door. Everyone stood up and began scurrying from behind the conference table heading for the exit.

Stepping into the corridor, each of the five NASA scientists were startled when they all were immediately encircled by ten special military security officers. Barry Harper stood back behind his security detail with two more officers on each side of him.

Once he saw the circle around John's team was fully completed, he looked to his right at one of the two next to him, and sharply said, "Turn it loose."

They both then turned and walked off. Barry then turned and walked through his line of men standing around the five stunned faces of the NASA crew up to where John was standing, and then shouted to his ten men, "To the mess hall for the magnificent!"

Walking down the corridor with the ensemble of security officers in the opposite direction of the two officers that Barry told to turn it loose, John walking alongside of him asked, "Why did you tell those other two officers to, 'turn it loose?' Turn what loose?"

"Press release," he quickly answered while propping his hand on John's opposite shoulder as they slowly walked, "Ordered directly by the White House after you and your team were initially briefed by yours truly, to finally let the population know that a new source of unlimited, incredible energy has not only been discovered, but harnessed by two of the top five scientists from NASA."

"I absolutely don't believe this," John said stuttering with a sheer bliss of astonishment.

"Your team John, especially Dr. Meyers and Dr. McKenzie, will now go down in history as the prophets that catapulted our civilization to the stars."

When Steven overheard that, he stopped dead in his tracks, but only for a moment with his eyes wide open along with his mouth once again. Barbara stepped up from behind and quickly thread her little arm inside his, and jerked him back into motion now walking together like a coupled escort.

The smiling ear to ear Jerry made his way up to walk alongside Miranda, tapped her with his elbow as he looped his arm up gesturing for her to slip her arm through his and walk together too as a coupled escort like Steven and Barbara.

Without moving her head, she just slowly swung her eyes toward him, and said, "I'd rather stick my arm through a meat grinder," and then shifted her pace a few steps ahead of him.

John then asked Barry, "Catapulted our civilization to the stars? What stars?"

Barry lifted his resting hand off John's shoulder a few inches to give him a couple of pats, and replied, "Any."

CHAPTER 6

THE ARCHIMEDES

All ten of the special team of security officers stood at attention while lined up on each side of the elegant, exclusive dining room's entrance hallway.

Five of the six figures sitting in the station's VIP dining room were barely nibbling on their elegant victory dinner celebration, for they were more concerned about through the recent course of events something much greater they were now trying to swallow.

From one end of the table sat John, with the Barry sitting directly across from him. To John's right was Miranda, then Barbara, Steven and Jerry.

At least for the time being, even the social flare of Jerry Danielson's reputation was idly at rest, with the same empty stare as his fellow teammates all glaring down into their plates fit for a king, knowing they were now headline news from one end of the Solar System to the other.

The sixth figure was the only one gorging away at his meal. He was the only one at the table without knowing the real feeling of what is was like to have your life just leap from one of daily routine to that of renowned recognition.

Barry hesitated his ongoing intake for a moment, and said through his muffled mouthful, "Hey! C'mon you guys. You're prominent public figures now! Illustrious legends! I have to go on living the life of Hubble VII's custodial corn dog, while you people…"

"So what happens now," John asked completely ignoring his remark, "I mean, where do we go from here with the, actually our, new discovery?"

Leaning down across the table to John, who was sitting directly on his opposite side, said with a heavy smiling sigh, "Truthfully, anywhere you guys want old buddy."

John swiftly snapped back, "That, is an epically vague answer Barry. You mind pinning that down a little more?"

Leaning back up, Barry quickly shot back, "In time John. In fact, a very short time. But for the present, first let me bring you up to par of what's transpired in the last two months since the success of that last test.

"Like I said before, they've been testing your discovery on the propulsion of unmanned probes. Five of them to be exact. Upon their return, the results each one revealed from their long test flights were not just identical, but so inconceivably extraordinary, it's almost too incomprehensible to believe. That's why they sent five, to be absolutely sure what they were looking at had actually taken place."

"Okay," John acknowledged, "What was it those probes found?"

Barry had just finished eating, and while slumping back in his chair, he answered, "It's what they found out about themselves, or rather what they found out they could do. Velocity. They achieved velocities so incredibly fast, that what I'm about to tell you, you might wish to take a moment to mentally prepare yourselves."

"Mentally prepared and ready for blast off," John answered.

Barry leaned in his chair to look directly in his eyes, and said, "Try almost eight thousand light years per second."

Silence once again shrouded over the five guests, just looking at Barry with a cold blank stare of disbelief.

After a few moments Jerry broke the silent shroud, "No way. The unlimited amount of thrust could maybe kick them to that speed, but what about the acceleration curve? Even as much as the acceleration shields incorporated into all the spacecrafts and unmanned probes we have today can withstand any amount of G's we throw at them, nothing would survive that much of a serious leap. Anything that could reach a velocity like that would be instantly disintegrated."

"Very true Doctor," Barry replied, "Oh, the first four probes never reached that much velocity, because their self defensive mechanisms would have shut them down instantly. However their instruments all calculated that much velocity could be achieved."

Miranda then asked, "Then what about the fifth one?"

Growing that same prideful smile again, Barry said, "Dr. McKenzie and Dr. Meyers' magnificent discovery can do much, much more than propel spacecrafts, my friends. The kinetic energy given off by the imploding tachyons can be so easily channeled to power just about anything. It's so unlike any energy source that's ever been utilized.

"So the fifth probe was designed with using that new power to not just get it up to that much speed, but to feed an equal relative amount into the newly engineered acceleration shield. Incredible stuff. It returned safely and completely intact, and according to the recorded data it made it up to seven thousand, six hundred sixty four light years per second."

John then began laughing, and remarked, "This just keeps getting better by the minute."

He then raised his glass of wine, and said, "Barbara, Steven, I guess I'm now compelled to give you both a little raise."

Everyone laughed as they raised their glasses, with a comment from Jerry after they all took their sip, "You two guys so really rock together. You should get married."

Barbara then began blushing, followed by a glazing new glare from her astounding beautiful eyes, and slowly resting the side of her face on his shoulder. Steven remained upright stiff as a board, with another frightful look in his eyes.

"C'mon Steven," Miranda said, "The least you can do is put your arm around her."

After gradually shaking his head, John said with a big smile, "I'm sure he feels the same, but needs a little time. Just always remember, you are one lucky young man Steven, to have Barbara's big heart."

"Here's to Barb and Steven," Miranda shouted as she raised her glass of wine, followed by the rest.

Lowering all their glasses, everyone got such a surprise when they heard the clapping of all ten security officers, also nodding their gestures of approval. Barbara was so impressed, she shouted thank you to them while waiving her little hand.

John asked Barry, "So have they performed any manned flights yet with the new source of energy?"

Barry rose to his feet, and replied, "This is the part I've been waiting for, my dear old friend. If everyone's ready, there's a shuttle waiting for us."

Everyone began rising, as John asked, "What's up now?"

"John," Barry said with a small giggle, "After everything that's happened, do you honestly think the United States Of America's only acknowledgement of gratitude for you and your team was this dinner?"

John now walking aside him simply said, "Never really gave it much thought, only because this has all happened so fast, like in the last three hours. Give me a little time to let this all sink in, and then I'll think about billing you."

Barry then said, "I have a feeling you're not going to bother with anything like that. I'm quite sure you'll be preoccupied with something else."

Gliding gently over the Hubble VII's opened dome, the shuttle began to drift into a downward slant, that everyone knew it would take them directly over the two magnificent eyes of both telescopes.

The two rims of the titanic eyes began to leisurely reveal their mighty presence, as the shuttle shifted her angle of approach to hover directly over them. The five scientists had seen it many times before, but it was such an extremely breathtaking wonder, that no one could ever get used to.

Both John and Barry stood back near the shuttle's rear viewport, both watching mankind's now second most magnificent achievement. The telescopes huge reflective mirrors were both a little over seven hundred feet in diameter.

The designer's original structure called for two telescopes, to make the view one incredible, superlative sight that constantly viewed the heavens like an immense pair of binoculars.

It was a government funded project, that the incredible magic ability of the designers was in fact well paid, but was never really given any public recognition for, strangely at the designer's request.

But now that those same five designers had discovered an incredible new source of an extremely powerful energy, that no doubt was going to revolutionize the entire civilization into the next advanced era of mankind's evolution. This was so overwhelming to every top representative in Washington D.C., and there was no possible way this special elite team consisting of five NASA scientists would go unrewarded.

"Attention everyone," Barry announced.

The rest of the team turned his way, and Barry began, "Almost two months ago, both the president and the entire congressional party unanimously voted to admirably reward the five of you with something that actually took over two weeks of debating up on Capitol Hill.

"When a decision was finally made, every space related scientist and space project engineer available were called in from every part of the Solar System to construct a very special instrument for the five of you, that will no doubt be very useful for your line of work."

The two massive cylinders of the telescope's outer chamber were now sinking below their line of sight.

Barry once again announced, "You might wish to look out of the forward viewport for something coming up anytime now."

The five of them moved up and stood just to the left of where the pilot sat, straining their eyes to find something from the shuttle's forward viewport.

After about ten seconds of craving curiosity, Jerry asked, "How's about a little hint as to what we're looking for?"

Barry directly replied, "You'll see soon enough Doctor."

The shuttle had slowed down a little, gracefully moving so close to one of the station's tremendous structural stanchions that you could see a few of the minor nicks in the exquisite streamlined flow of high alloy steel, since the station's magnificent birth three years earlier.

"Hang on guys," Barbara swiftly said, "I see something. Look, just too the left side of the upper cross section stanchion. See it? Anybody?"

"Barbara," Steven said, and then leaned over and whispered in her ear.

She then looked again, straining her eyes even more, while repeating back to Steven, "Are you sure?"

"Care to share any of Steven's advice Barb," Miranda asked.

The rest of them were now closing in on her even more, almost to the point of smashing her up against the glass of the viewport.

With the exception of John, who now stood right behind the other four, that still had their faces almost clued to the glass, Barbara then confessed, "Okay, Steven. You're right, it's only the leg for the emergency transmitter…"

Suddenly the four of them instantly jumped back, slamming into John, followed by Jerry's sharp, "Whoa!"

The bottom of the front viewport began to rapidly fill in upward by something just on the other side of the viewing port's glass, matching precisely the same rapid speed the shuttle was drifting in a downward motion.

With his now ear to ear smile, Barry bent over to instruct the pilot again. The shuttle then began to back away from what the viewport was now slowly beginning to reveal, and with the overtaking shroud now far enough away, the vast shadowed silhouette began to take on a somewhat familiar shape.

Developing into a more solid contoured structure, the five shocked NASA faces could now tell it was some type of magnificent spacecraft. It stretched out all the way from one of the station's massive foundation stanchions to another, that everyone onboard knew the stanchions measured a little over a hundred meters apart.

Barry leaned down once again to say something to the pilot. The shuttle then began thrusting forward, accelerating to bring it up even with the massive ship's front section, and then maneuvering it around to give John and his team a stem to stern view as it slowed down just enough so everyone could get a full view as to what they were now looking at, stunned with utter bliss.

John was the first to step out of his shock long enough to ask, "Barry, are we looking at what we all think we're looking at?"

The station commander smiled again as he replied, "She's a beauty, eh John?"

To the five NASA scientists, beauty was an understatement. The structure plating edifice used in the construction of the ship's striking color was so very black, it actually had a smooth blue glazing shimmer to it.

The bow section was sculpted in a cone shaped conduit fashion, spreading outward from the bow's center point to join the long streamlined fuselage that extended over a hundred meters all the way back to its rear stern. There, two majestically curved fins protruded upward from each side, that were separated by two tremendous silver cylinders that lay side by side between them; the ship's thrust exhaust ports.

Barry stepped up between the viewport glass and the five speechless NASA scientists, and announced, "There it is, people. The fruit of your labor, named after an ancient Greek architect that I'm sure would be proud to know the five of you. Behold my friends, the *Archimedes*."

CHAPTER 7

DIRECTION

The shuttle pulled away from the docking platform, leaving behind the station's commander and five mesmerized slow staggering scientists, all looking up to the right with their eyes and mouths wide open.

Barry then said, "C'mon, let's pick it up a little. There's a ranking official onboard that's waiting for us. It's his turn to answer the rest of your questions."

After walking throughout Barry's main tour of the entire ship, the six of them all stepped through the threshold door to the vessel's main bridge. Looking around at all the instrument panels and new equipment, Jerry more mumbled than spoke, "Whoa. Wonder who's supposed to fly this thing?"

A familiar voice then emerged out from back behind them, "You are, Jerry."

They all spun quickly around, just in time to see a tall, middle aged military man with glittering stars on each shoulder and two guards on each side of him, step up to Jerry and shake his hand proudly, and say, "It's always an honor to see you Jerry. Always."

Jerry replied with an unknown sad sense of recognition, "Thanks. You too General."

General Vaughan deeply sighed, and then turned to John, and said, "You're looking good John," and then turned to Miranda and formally said, "Dr. Young."

Miranda nodded to him saying only his name as

well, and turned to the two frozen stares of astonishment, and said, "Dr. McKenzie, Dr. Meyers, please allow me to be the first to beg you both for forgiveness regarding my previous staff's misappropriated carelessness regarding your extraordinary discovery. I can only hope our urgent haste to construct this magnificent vessel for you and your team will at least gratify a little of our nation's immeasurable lack of indifferent disregard for your magnificent, historical breakthrough."

Miranda quickly replied, "Barb, Steven, what he's trying to say is that he's sorry for his office's magnificent screw up."

One of the stoned faced guards quickly burst out with a quick spurt of laughter, then immediately resumed his serious stance with a touch of worry gleaming from his face.

General Vaughan swiftly turned to them and ordered them to wait out in the hall and to shut the door.

He then turned to John, and remarked, "She actually said it better than I. Thank you, Dr. Young, but I believe we have more than well compensated for our, 'screw up', and Jerry, this vessel is was actually designed to handle easier than anything you've ever flown. So, where do you people think you want to go first?"

Everyone just looked at each other in amazement, until John asked, "Just how far can it take us?"

"Anywhere in the universe," he proudly answered, "Or the next. This is incredible ship is your home now. That is if you want it to be."

John looked at his shocked staff, and then replied, "Mind if we discuss this matter alone first?"

46

He started to walk for the exit, stopped a moment to pat John on the shoulder, and said, "Take your time, because time is the one thing your incredible discovery has solved, because wherever you and your team decide to travel, it won't take hardly any time at all to get there."

General Vaughan then turned to Jerry, and sadly said, "Take care Jerry. See you soon."

Jerry replied, "Thanks. You too. One question. Just how fast can this thing go General?"

The General smiled, and then replied, "Seven thousand, six hundred and sixty four light years per second. Just let me know when you're leaving. The destination is your call guys."

He then shut the door, leaving five NASA completely paralyzed, only shifting their eyes to one another.

After a few moments, John instructed his crew, "Get your affairs in line, and when we all get back we'll discuss the next move. Go."

Three days later they all sat back in the same place, the main bridge, all looking outward toward the stars.

"Okay," John began, "Any ideas as to our first destination?"

"Doesn't matter to me," Jerry was the first to respond, "I took this baby out for a spin, and the General couldn't have been more correct. This bad boy can take us anywhere. So you guys decide, and then all you have to do is point. I'll take it from there."

Miranda then said, "Should be your call sir. You're the boss."

"This should be a democracy Miranda," he shot back, "Barbara, Steven? This was your invention, and we wouldn't all be here if not for you two."

Barbara then suggested, "Steven and I have been discussing this matter. We vote to go all the way sir."

"Exactly where's that," John asked.

Steve then said, "All the way. Sir."

"That's extremely vague junior," Jerry added, "Care to be a little more specific?"

"All the way to the end of our universe Jerry," Steven answered, "Barbara and I have calculated that at this vessel's maximum velocity, it will take three weeks to travel thirteen point seven billion light years. We would really like to see this, "Big Bang," which she and I don't believe really happened. Let's go and see what really took place."

Jerry sighed, and then remarked, "That's a long way from home junior. What if it really did happen? We'll be blown to kingdom come!"

Jerry then turned to glance at John's reaction, but something was in the way; Miranda's finger pointing directly at his nose.

With his eyes crossed looking directly at the end of her finger, she shouted, "You call him junior one more time, and I'll break your face! He does have a name you know!"

John shouted, "Okay, as you where Miranda! Back off, now!"

John sighed in frustration for a moment, and then began pacing, and then said, "I'd like to know as well guys. But anywhere we go, I'm making this clear right now. It must be unanimous."

"But boss," Jerry said, "Do you know how long it would take just to get out of our galaxy?"

"Not if we take the path of least resistance," Barbara told him, "All we have to do is shoot straight up above the spiral arm we're in of the Milky Way to a safe enough distance out of it, then straighten out and sail directly over the clear space above the Milky Way in the direction of Andromeda."

"She's right," Jerry said, "That never occurred to me. We would avoid transiting through any galactic debris that we could possibly get in our way or run into."

"Then you have my vote," John said, "Besides, if it wasn't for you and Barbara, we wouldn't be even talking about this. How about it? Miranda, what do you say?"

She replied instantly, "I say they deserve it. No problem here."

A shroud of silence then engulfed the bridge, as everyone looked at each other with cheerful disbelief, unable to bring themselves into the concept that it was actually about to happen.

Miranda then broke the ice, and shouted with a tone of humor, "Well, what are we waiting for? Let's go!"

"Let's rock, guys," Jerry agreed, "There's a whole universe out there waiting for us!"

CHAPTER 8

LIFT OFF

"Steve, Barb? How far do you think we need to get away from here before we head upward," Jerry asked.

"Doesn't really matter," Barbara replied, "Anywhere really. Your call, Captain Danielson!"

"Roger dodger Barb," Jerry said, "So fasten your seat belts, and make sure your tray tables are up and in the locked position people. Here we go!"

"The new acceleration curve mechanism is online," John said, "Right Jerry?"

"That's affirmative boss," he answered, "Comes online automatically when the artificial gravity does."

Miranda couldn't resist, "Which you did too, right?"

"All but yours sweetheart," he snapped back, "Banking the bow ninety degrees up now."

The star field ahead dropped quickly down from the large front view window, and three seconds later Jerry shouted, "Instruments say all is well! Bringing main thrusters online in five, four, three, two, one, lift off!"

The powerful tachyon powered engines began to rumble with a serene smooth silence, barely shaking the **Archimedes** floor decks to a mild tremble. The few dots of light they could see from the stars ahead began to gradually move toward them.

John said from his strapped in copilot seat, "Still can't fully believe this is happening yet. How about the rest of you?"

Barbara shouted from her seat behind them with Steven on her right and Miranda on her left, "Still pinching myself, making sure this is no dream sir, and that gives me the excuse every time I pinch Steven too!"

She then leaned over and kissed him on the cheek, and he just turned red with a bigger grin each time she showed her affection for him.

Miranda then followed with, "Alright you two! Behave yourselves for a while! This is history, and we're making it!"

Barbara laughed as she shouted back, "You're just jealous because you can't do the same with Jerry at the moment!"

"Oh, trust me girl," she replied, "He has nothing to worry about that coming from me! How we doing up there sir?"

"Don't really know," he said back, "Let's ask our pilot. Jerry, how fast are we moving now?"

"We're now, let's see boss, two thousand, seventy one light years from when we first took off, which was forty nine seconds ago! Can you believe this boss man?"

"Not yet still Jerry," he shouted back, "Okay! Time for the first order of business into our flight time. No more, 'boss,' or 'sir,'! I want to be just John to my 'family' now. I really love you guys as if we were really blood related. But since we're not, that comes directly from my heart! Okay?"

Steven spoke for the first time in a while, "That's going to take some getting used to sir! I mean, um, John. It's just that we respect you so much, that's going to be a hard one for me."

"I know how you could solve that one Steven," Barbara shouted, "Just call him Dad!"

Everyone burst out laughing, John Stanley the loudest.

After their euphoric states that brought the laughter on faded down, John asked, "Captain Danielson! How much further do we have to climb above the Milky Way before we straighten back out and head for eternity?"

"I was just getting ready to ask Admirals' Meyers and McKenzie. We're, let's see, three thousand eight hundred forty five light years up guys, and our speed is still increasing. Your call, Admirals!"

Barbara and Steven spoke briefly, and then Steven shouted, "Ten thousand light years should be fine, Captain Jerry!"

"Setting course to turn us at that distance now, Sir Steve! Done!"

Miranda shouted even louder, "I got a good question for everyone! Why are we shouting at each other?"

Barbara quickly shouted, "I'm so excited I could scream the rest of the way, Mir! It's exactly what you said about us making history, and can the rest of you imagine how much in all our futures we're going to be thinking about this very moment?"

"Thanks to you two lovebirds," Jerry said with pride, "C'mon, Steve, pinch her back to see if she's dreaming!"

Barbara then screamed, "Ouch! He actually did it! I'm so proud of you even more now!"

Everyone burst out laughing again, until John said, "Let's see the view behind us of home, guys. Let's see here. Ah! Okay, the monitor above the window should show, there! My God, is the Milky Way really that incredibly beautiful? Look at that!"

"Oh my God," Miranda softly said, "Look at those enormous spiral arms guys. That's the most spectacular thing I've ever seen!"

Jerry added, "Oh, trust me Miranda, I'll bet the farm this trip that we're going to see a lot more of God's creations even more spectacular."

"Jerry," Miranda said softly back, "For once I'll agree with you. Which arm is our home in Jerry?"

"The Orion arm," he answered, "But I don't know which one it is."

"This should tell us," John said, "Watch for the little yellow plus icon marker. There you go."

The true magnificence of the streamlined glazing light-blue depth of the home spiral arm mankind evolved from took everyone's breath away. The silence of that magnificent moment they were all captivated by was no estimated painting, which could never do the true real sight of where the human race originated any justice whatsoever.

"All my life," John said with true emotional words, "I've always wondered if mankind would ever get this far. And now here I am with the only family I've really ever had actually looking at it."

"That's from a monitor John," Jerry said with stuttering softness, "Watch this. Ten thousand light years in three, two, one. Here goes."

The *Archimedes* began to shift course to a level plane, and the true sight of the Milky Way crept its way up from the bottom of the huge front view window. Jerry turned off the internal lights inside the ship's bridge, and Barbara's sweet crying voice flowed its way throughout the breathtaking beauty the darkness now truly revealed of the home galaxy that the civilization called mankind came into existence.

Miranda said with a mumbling voice of tearful tone, "Jerry, you're right again. We are in for one hell of a journey. Look at that. Who would ever believe that almost fourteen billion years of evolution could bring this into unbelievable reality."

"You know what the greatest part is that we're looking at Miranda," Jerry said with a touch of cheer, "You, me, and the rest of our family, here onboard and down there, are all part of that same incredible creation of astonishing beauty. And now God has granted us the privilege and honor to actually see it for ourselves."

She unbuckled her seat straps, got up and walked over to where he sat, leaned down and actually hugged him with one arm, and then softly said, "You're right again, Captain Danielson. Maybe, just maybe you're not so bad after all. But I doubt that."

Without looking back up to her, he smiled with such a leap of joy, as he replied, "I have moments, Miranda."

John had already unhooked his chair straps, and stepped over and hugged Miranda with true pride as he embraced her, and then said, "From what I just saw, I now believe in God more now than I ever have."

Miranda slowly smiled from ear to ear, and placed one finger over her lips gesturing for silence, and then pointed toward the front window.

John slowly turned around and didn't yet notice what she was motioning to.

Barbara and Steven standing side by side right up to the glass, with Steven's arm finally around her, and their heads leaning against each other looking at what their discovery had now treasured them with.

God's greatest creation of beauty, so much more powerful than any magnificent view of His magnificent work of the universe that He made billions of years ago. The most immeasurable, impregnable force of His most powerful ability.

The love that Barbara and Steven now held for each other's heart and soul.

Jerry then rose from his helm control post, walked over and put his arms around them, and lowered his head down to where his chin almost touched their faces.

He then whispered, "Sorry to interrupt, my little brother and sister. I just need to know the direction you wish me to send us. Take your time though, and enjoy this moment of the greatest victory in the history of our race. And thanks so much for giving me the honor of being part of your triumphant success of mankind's greatest conquest from your astonishing, accomplished achievement."

Steven just glanced at him and winked, and Jerry slowly let go and walked back over and stood next to Miranda and John.

Miranda then gently whispered, "God does some pretty great work guys."

John followed with, "Sure does Miranda. I had no idea it would be this fantastic. I can't begin to imagine what we're going to see on this historical journey."

Jerry leaned down in front of Miranda, and quietly replied to John, "I'm sure that's very true, but I don't think that's what's she's referring to."

He then sat back down in his pilot's chair, and preset the instruments for an upcoming course correction.

John looked at Miranda somewhat puzzled. He then glanced back toward the window, and smiled when he looked back at Miranda, knowing then what Jerry and she originally meant. She slowly shook her head up and down, knowing what John just realized.

She slapped Jerry on the back, and then said to John, "Sorry. I didn't want to admit he was right again."

John burst out laughing, and as he turned to sit back down in the copilot's seat, he looked back up to Miranda and said, "He does have more moments than he'll ever admit to Miranda."

"We'll see sir," she said, and then asked, "Okay people, since its celebration time, what do you guys want me to make from the galley?"

Everyone then exploded laughing, even Miranda, when Jerry remarked, "Why don't you just stay there and we'll let you know, because that's where women really belong anyway."

"I'll surprise you people with something good," Miranda said, and then slapped Jerry on top of his head as she turned to walk to the galley, and looked toward John still laughing a little, and said, "Like I said before, we'll see."

CHAPTER 9

JOURNEY'S HEART

The *Archimedes* was now soaring over the Andromeda galaxy, using it and the direction line toward the Milky Way as a homeward beacon, in case they were to somehow get lost. This way they would just stay on this straight line course until they got to the thirteen point seven billion year destination.

Everyone was so stuffed from the delicious Mexican meal Miranda made, and only Jerry and John remained now on the bridge. John's head was now bobbing up and down a little from the great celebration meal knocking him and everyone else out cold. But he couldn't leave the magnificent view before them.

Jerry was stretched out into the pilot's seat that reclined back enough to almost a level position, and had the navigational computer on auto pilot. But his eyes were wide open, thinking about something that really had him upset from time to time.

John turned over to lay on his left side, but conscious enough to notice Jerry was still wide awake.

"Hey Captain," John said, "Why don't you go to your quarters and get some rest Jerry. I'll take it over for a while. You've really been grinding hard for almost three days now for this trip. You've earned some real shuteye time."

Jerry sighed, and then said, "I'm good boss. I mean John. Just mentally drifting around the universe. It's so peaceful out there."

"You sure guy," John answered, "Wait, I know. It's all your women you miss. They'll still be there in two or three months when we get back. I mean you really attract them like nothing I've ever seen. You even have a, 'Dr. Jerry Danielson Fan Club', that's one hundred percent beautiful women."

Jerry snapped out with a surprising tone of anger, "It was never my idea to have that stupid club. Just a bunch of crazy women put that together. To tell you the truth, and please don't repeat this, it's really annoying as hell."

"Hold on Jerry! Don't get mad at me for telling you something you already know. You're really one good looking man Jerry. How can all those beautiful women annoy you? Men dream of things like that. If it was me, I'd be proud."

"I'll admit it's flattering. But John, there came a day in my life when I realized I wanted to be really loved by one good woman, for just being me. What I really mean is not for my looks, but for who I really am inside."

"Wow, Jerry! This is a part of you I never dreamed you had some kind of problem with. Does anyone else know how you really feel?"

"Nope. Well, one person does now. You. But please don't ever repeat this to anyone, okay?"

"I thought you knew me well enough to know I would never break a trust for someone I admire as much as you. Man, Jerry, you should have said something to me sooner. I've always not only considered you as a good worker I could always depend on, but also as a good friend."

Jerry slowly smiled, and said, "Now that's a part of you that I never knew about!"

They both began to laugh a little, and then Jerry asked, "Why didn't you ever tell me boss? I mean, well, you know what I mean."

"How do you think the other three would feel about that, or the NASA Board of Directors, or even General Vaughan? They would consider that very unprofessional. Even knowing you're my number two man."

"Yeah, you're right. I wouldn't have said anything either. Sorry."

"Well, at least none of those Jerry hungry women can get to you out here."

"I don't know boss, some of them are pretty persistent."

John started laughing out loud, while he stretched out his chair as well, and then said, "That's one of the reasons I always wanted to be your friend Jerry, because you could always make me laugh more than anyone ever could. You've got the most gifted wit I ever heard."

"Thanks, John. Coming from you, that really makes me feel good. But having the ability to get almost any woman you want is not what you think it is. There are some women that literally hate me! I'll give you a perfect example. Miranda. What have I ever done to make her hate me the way she does?"

John stuck his hand up for Jerry to stop, and said, "Do you really think I'm that blind, Jerry? There's been times I've wanted to fire her for the way she treats you. It's just that I've always thought you never really cared, because of all the other women that all you would ever have to do is tap your com button signal, and call any one of them, and they would be delivered to you like ordering pizza."

Jerry just smiled, and said, "You know, it was even simpler than that. But as far as Miranda, and oh God if she knew this I would just blow myself out of one of the airlocks. I like her John."

"Don't see how. What upsets me about that is sometimes it seems like she goes out of her way to get on your case. But my instincts have always told me to just stay out of it, because you never reduce yourself down to her level by retaliation. You always keep your cool, knowing it will pass."

Jerry just giggled, and then said, "Thanks. You're one cool boss, boss, but I'm afraid it's not that simple. Let me try this once more, and we'll never bring it up again. I really, really like her. Much, much more than a co-worker should have for a colleague. Okay? Now, case dismissed."

Jerry just sighed heavily, and then turned to look out at the splendor of millions of white glazing dots coming toward them. Except these were not stars. What now laid before them were entire galaxies approaching them.

He then turned back to John, who was now sitting straight up in his chair, looking straight at Jerry with a look of shock, but with an ear to ear smile.

Jerry instantly grew a big smile as well, and then began to laugh as he asked John, "What?"

"No way," John replied, "C'mon Jerry. Are you serious? Really serious?"

He turned back to look out of the front view window again, and softly said, "From the very first time I ever laid eyes on her. The day I met her, that, is the day in my life when I realized I wanted to be really loved by one good woman. Her of all people. That's my luck, probably for sinning with all the other ones that I really didn't care about, and I've broken so many hearts that it got to the point of not really caring anymore. Then she comes along, and not only breaks my heart, she thinks I'm some kind of women vacuum cleaner that hops in bed with every skirt that looks my way!"

John laid back in his chair, unable to stop his mild laughing, and finally said, "I can't believe this. I will admit, she is one of the most gorgeous women I've ever known, but this takes the cake. All of it. You're really in love with her?"

"Head over heels. It's so strange, because I've had mild crushes and a few infatuations, but nothing, and I do mean nothing ever like this.

"Now, you have to promise, although I know you won't ever say anything, but for my emotional security anyway I still want to hear you say this out loud, and will always stay between us. C'mon John, please, or I'll have insecure nightmares if you don't. Now, try this one on! I haven't been with another woman since the day I met her!"

John began laughing even harder now, but did regain enough control for a moment to say, "No, I will never say a word, but that's been over three years now!"

"Thanks," Jerry replied, and began laughing along with him, and stuttered out, "But if you keep laughing any harder you're going to break something inside, and don't expect any CPR from me! You're not my type!"

They both then began to lose it completely, and after about five minutes were finally able to regain control, and Jerry then asked trying to catch his breath, "Okay, okay! What about you, mister laughing man? The guys and I know nothing about your personal life. Nothing. We just know you're not married. Do you have someone special, or even any girl friends, or someone that you date?"

John then calmed back down somewhat and ultimately was able to say, "Not really. In fact no one at all. Too married to my job I guess. I was married once though."

"Really? You're pretty much a woman head turner yourself John, and every time I've ever seen a woman trying to move on you, you're all business and you somehow manage to scurry away. Does it have anything to do with why your marriage didn't work out for you? Care to talk about it? I'm a good listener."

Jerry then became instantly shocked, when he heard John spurt out a small laugh.

"I heard that! Okay, now it's your turn boss! Let's hear about it!"

John just shook his head for a moment, sighed deeply, and then said, "Came home sick at lunch one day, and caught her in bed with someone else."

He then spurt another short burst of laughter, and Jerry's face froze with shock.

Jerry then shouted with a serious growl, "Now you're really scaring me. You came home and caught your wife in bed with another man, and you think that's funny? That's damn dysfunctional. I mean really. That was always one of my greatest fears, and you're laughing about it?"

John tried to hold it back, but couldn't help smiling, and said, "Alright. But this stays between the two of us, because I've never discussed this with anyone. Agreed?"

"My Lord," Jerry replied, "And now you're actually grinning again. Holy smokes, I can't wait to hear this one, because it can't possibly be worse than what I've been going through for three over years now! Of course I'll never say anything, because look what you have over me now!"

"Oh, you'd be surprised just how things can get you so twisted up inside when it comes a woman you love my good friend, far worse than what you're going through, because what happened with my ex-wife caused me to develop a serious phobia from ever getting involved with another woman again."

John then looked away spurting out small bursts of laughter again.

Jerry then shouted to the top of his voice, laughing between each word, "Because of always seeing the image of her in bed with another man? Christ, John, if it's making you laugh, it couldn't have been that traumatic! Maybe for a while, but c'mon boss, move on! How on earth can that give you a lifetime irrational fear of ever being with a woman again?"

Still laughing a little, John calmly said, "That, is not the accurate appraisal of what I said Jerry. You apparently were not fully listening. So, let's try this once more, and in your own words, case dismissed.

"I said, I came home one day and caught my wife in bed with someone else. When I came home that day, I didn't catch her in bed with another man. I caught my wife in bed with another woman."

The look on Jerry's face turned completely paralyzed, to a staggering state of catatonic cold stone dysfunctional withdraw. He gradually turned to look out the front window, feeling an oncoming surge of hysteria building up inside him, until something else to the right of him swung his attention back; John standing next to him with both arms leaning down to look Jerry directly in the face.

John was mildly laughing, and finally was able to say, "And not just any woman! Another woman the size of an elephant!"

That was as much as Jerry's conscious withdraw could take, and he and John now looking right into each other's eyes, both exploded into the most uncontrollable hysterics any of the two had ever experienced.

Screaming so loud now with convulsive laughter, Jerry lost all control of his motor functions, bending over and then sliding down his chair. John was almost as worse, with his head turned sideways pressed into Jerry's back, as Jerry finally slid off and dropped to the floor, impacting the deck on his entire left side.

John managed not to fall into Jerry's chair, and lifted himself up and turned toward Jerry's spastic state, who had rolled to his side now laying on his back. His legs were now spinning in the air like he was riding a bicycle, holding his stomach with both hands.

Staggering over to Jerry's screaming, out of control state of laughter, John stood over him with the same river of tears pouring down his face from his own uncontrollable screams of hilarity.

He softly kicked Jerry in the side, shouting through his laughing, "Still think your issue is worse? Shut up, you squealing moron!"

John tapped him again in the side with his foot, and screamed again down to floor, "Shut up, before I pee on myself," but then heard something in the rear of the bridge.

He looked up for a moment, only to see Steven's shocked figure standing in the rear with one hand holding the bridge door open, and then slamming it shut as he ran away outside the door.

John then dropped himself into Jerry's chair, and started falling back into hysterics again, as Jerry rose himself up enough to see what the loud slam was about.

Through his hard breathing, he asked John, "What was that?"

John opened up his hands enough that now covered his face, took a pausing deep breath, and quickly said, "The terrified face of poor Steven! I think we're in real trouble now!"

Jerry's back collided down to the deck again, whaling with unrestrained wild laughter once more, shouting when he could, "Poor kid, oh that poor kid!"

John swiftly regained his control, stood up and held his hand down to Jerry, and exclaimed, "C'mon! Grab my hand, quick!"

He pulled Jerry's limp body off the floor, and sat him back down in the pilot's chair. He then stepped back, and leaned against the rear window, rapidly wiping his soaked face off with his handkerchief.

Tossing it quickly over to Jerry, he said, "Here, hurry up and wipe your face off. If my guess is right, Miranda and Barbara are going to come charging through that door any minute now, and I'm feeling too great now to be a victim of mutiny this first historical deep space flight! Hurry up!"

Jerry rose up from the chair as he got the last of the tears dried up, walked over and leaned his back against the window standing right next to John, both now facing to the rear of the bridge.

Handing the handkerchief back to John, in his exhausted state, Jerry said, "Thank God you had the willpower to come to your senses fast enough."

John yawned out in relief, and then said, "It's just that those two ladies are so protective of him, there's no telling what he told them, and I'm sure by now they're well on their way here."

Suddenly the rear door blasted wide open, with two women that had their eyebrows in a complete dagger shaped posture, slamming the door behind them swiftly walking up to John and Jerry's looks of mild confusion.

Miranda and Barbara stood right before them, now with a little more calmer glaze about them, as John asked, "Yes ladies? Something wrong?"

They looked again at each other, and as Miranda looked back to them, Jerry asked, "What's up?"

She took another step toward them, and then asked, "What in the hell is going on in here?"

The two men looked at each other like they were a little confused, and Jerry mildly asked, "Going on in where? What are you talking about?"

Barbara by then was right beside her guardian angel Miranda, and said with a little anger in her tone, "Steven's so shaken up right now we could hardly understand him. About five minutes ago I heard someone running down the hall by our quarters, and then slam the door right next to my room where Steven's room is.

"I went over, and he's on his bed in the fetal position with his pillow tucked under him, saying Dr. Stanley's gone mad, and he's on the bridge kicking Jerry to death!"

John looked at Jerry, smiled and then said to Barbara, "We were just goofing off about something, and laughing about it. That's all. I mean I'm fine."

"What it was Barb," Jerry said, "Is that I tripped and fell, and boss man here was standing over me trying to help me back up, but we were laughing so hard at me falling because I get so spastic sometimes. That's all."

She looked at him funny, then to John, and grew a big smile, and said as she turned to leave, "Yeah, right. Goodnight everyone."

She opened the door to leave, and Jerry shouted, "He just needs some cuddling on, Barb. So go and smooch on him a little."

"I intend to," she answered, "But the next time you guys get goofy when you're really talking about women, try and leave him out of it. Goodnight."

Miranda stood firm, shifting her magnificent beautiful eyes back and forth to them. She turned to leave and remarked, "According to Steven, that's not what he said he saw. Goodnight."

Walking toward the door, Jerry shouted to her, "He just needs some loving on from Barbara, and he'll be fine."

She looked back to Jerry, and said, "Typical advice coming from a playboy like you."

She shut the door, and Jerry just grinned. He then looked over to John, and said, "Think I'll take you up on your offer to hit the sack, because after a laugh like that, I've lost about ten pounds."

"Sure Jerry. I hope what she just said didn't ruin your mood. I really needed that. Think you did too."

"Oh, John, after what you told me about your predicament, and the way she treats me, I'm going to throw in the towel on women. But Oh God thank you boss. I really needed that laugh. Goodnight."

"Goodnight Jerry. But can I ask you something first? Have you ever considered just telling her how you feel, and have always felt? I mean, what have you got to lose? She just might feel the same way about you, and treats you like that because you won't ever move on her."

"I have another idea, John. Thanks to you. Every time I start thinking about her, I'll just fantasize coming home, and catching her in bed with Bimbo The Hippo!"

"Oh, trust me Jerry, if that were to ever happen to you, you'd never be the same, and not just about her. You'd feel that way about any woman. Goodnight."

CHAPTER 10

THROAT OF DEATH

General Vaughan had gone all out to reward his five NASA heroes. The view from the luxurious observation deck on the top of the **Archimedes,** positioned directly over the main bridge was absolutely breathtaking.

The room was a little over three thousand square feet, equipped with three plush sofas and three huge recliner seats all lined up against the gradually curving rear wall, set up to where the viewers could observe the downward curve of glass dome that went all the way down to the front side's floor, and the entire room could be adjusted to be as softly lit as possible.

The elevator doors silently slid open, and Miranda walked out and over back to one of the sofas and fell face first into it. Steven was sitting over on the other end in a recliner, speed reading his way through one of the many books he brought.

She then rose up to make sure it was him and tell him hello. Watching him flip through each page every three seconds always fascinated her.

"Hey, Steven," she said.
"Hello, Dr. Young."
"How you doing?"
"I'm fine. Thank you."
"Doing your reading?"
"Ah Hum."
"Can you really read that fast, Steven?"

"Ah Hum."

She finally rolled her eyes up with a touch of frustration knowing he hadn't heard a word she said. Miranda just looked up at the unfathomable view of the amazing splendor of the vast ocean of magnificent beauty slowly drifting by them.

Millions of majestic shaped galaxies, with a breathtaking rich glaze of such unusual superlative colors rising from their titanic swirling configurations.

Whenever the heavy distant groups of the approaching galaxies would get near enough, they would then gradually begin to spread open, as if they were commanded to move aside to grant safe passage to them.

In her rapture she said in a soft, beautiful tone, "All I ask is a tall ship, and a star to steer by."

Steven replied, "Ah Hum."

"I've always loved that quote Steven," she softly said to him, "It's always been one of my favorites. I just can't remember who wrote it. Do you?"

All she heard this time was pages swishing every few seconds, with no reply at all.

She just smiled this time as she slowly shook her head, until surprisingly another soft toned voice came out from another sofa on the other side of the room, that said, "John Masefield."

It was Jerry's voice, apparently stretched out on another one of the sofas's that she didn't notice when she came in the room.

"Jerry? I didn't know you were here."

A few seconds later through the serene darkness, he

said, "Ah Hum. That's always been one of my favorites too."

"Didn't know you were in to poetic quotes."

He softly replied, "I must go down to the seas again, to the lonely sea and the sky,
And all I ask is a tall ship, and a star to steer her by.
And the wheel's kick and the wind's song and the white sail's shaking,
And a grey mist on the sea's face, and a grey dawn breaking."

Miranda slowly rose up and looked around for him, but lying like she was there was no way to tell. So as she slid back down, she remarked, "That's beautiful, Jerry. Thank you. You do have moments."

"Actually that's only the first paragraph from his three paragraph poem, "Sea Fever," that he wrote over six hundred years ago. Yet, the beauty of it Miranda will remain forever, just like what we're looking at now."

"What we're looking at now won't," she said, "Because the universe will eventually burn itself out one day."

After a few calm seconds, Jerry said back, "The beauty that we know from splendors like this will live in our hearts for eternity."

She rose up again, and said, "That's even more beautiful Jerry. Is this one of the ways you charm all your girlfriends?"

After a few moments of silence, Jerry sadly said, "I don't have any girlfriends Miranda."

She couldn't believe what she just heard, and after a few moments of thought in the smooth silence, she said, "Not that I think I'm getting space gaga, but did I just hear what I thought I heard?"

"There's nothing wrong with your hearing, and we're all getting space gaga. We've been gone over two weeks now, with only a week left until we get to the outer edge, and we're all getting a bit space happy. Here, I'll give you a perfect example."

He then shouted, "Steven?"

"Ah Hum?"

"The ships burning up, and we're all going to die."

After two paper swishes, they heard, "Ah Hum."

"See what I mean?"

She mildly started laughing, followed by him for a few minutes.

After they calmed back down, she said, "You do have moments Jerry. Thanks, I needed that one."

"No charge."

She then replied, "Well then how about out of all your non-commitment flings?"

"I don't see anyone. Haven't for some time now."

"What," she said louder, "You've got to be kidding me. The man that has a fan club, all women, doesn't see any one of them?"

"That's right. And if it makes you feel any better, I've never been to that club. Get a lot of mail, but I've never opened one letter. It still keeps coming for some reason, but I haven't seen or been with any women for a little over three years now."

Miranda hesitated for a moment, and then said, "I can't believe what I'm hearing. For over three years, you haven't been with any women? Not even a friendly date?"

"Nope, no one. No one at all."

"I find that a little hard to believe Jerry. But if that's really true, any reason why?"

"Sure is," he answered.

"This ought to be a good one. And the reason is? Jerry?"

Something else had caught his attention. She then heard, "My God, look at that monster!"

She quickly looked over her shoulder, and rose to her feet.

Coming up slowly on the ship's left side, was a galaxy so huge it was almost inconceivable. Tens of thousands times bigger than the Milky Way.

Miranda shouted, "Steven! Look at this!"

Steven just answered, "Ah Hum."

Jerry then shouted to him, "Steven, take five on the reading, and look at this. Now!"

Steven then walked up to where they were standing, and said, "I can't believe how big that thing is."

Jerry then shouted, "It's not just that I'm talking about! The gravitation pull coming from the supermassive black hole in its center could suck us in, even from this distance!"

Steven calmly said, "Well, I don't know about from this distance, but…"

Jerry shouted to them, "C'mon!"

The three of them then began walking quickly to the elevator. Jerry slapped his com button, and shouted, "John! You on the bridge!"

"Keeping your chair warm, Jerry. What's up?"

"That monster coming up in our eleven o'clock position. We need to change course immediately and swing around it."

John replied, "I was considering it, but I believe were at a safe enough distance from any danger from it."

They stepped into the elevator, and as Miranda shouted to the voice activated control for Barbara to come to the bridge, Jerry was shouting, "We can't take that chance boss! We don't know enough about black holes that size yet to assume anything! I'll be there in a second, but you need to change course now!"

The elevator swung open, and as the three of hurried out, Jerry shouted as he ran up to John sitting in the pilot's seat, "Alright boss, I'll take it from here!"

John just sat there shaking his head and calmly said with Jerry standing over him, "Calm down, Jerry. Look at the gravitational detection gauge indicator. It doesn't show any abnormal outside tugs, or any coming toward us."

Jerry swiftly snapped back, "John, the Hubble VII's team showed over a hundred years ago that even an average size or wimp black holes cast out unpredictable waves of enormous gravitational waves from the center of the vortex. It's got something to do with it having to cast off excess amounts of energy intermittently, when it tries to swallow too many large celestial objects at one time, like huge star clusters or even small galaxy groups!"

John snapped back, "Jerry, the gravitational detection gauge would warn us of any approaching waves of that type. So would you cool it with the unnecessary theatrics?"

"Boss," Jerry shouted directly in his face, "The vortex in that monster is bigger than the entire circumference of the solar system, which means it eats galaxies the size of ours like we eat peanuts! Those immense erratic waves are too unpredictable when they're released, and could reach us in a matter of minutes! I need to put as much space between us and that thing now! So would you please get up and let me do my job!"

Barbara had already swung the bridge door open and trotted up to stand behind Miranda and Steven, just in time to hear what Jerry was screaming at John about.

Before John could say anything back to him, Barbara had already seen what was gradually coming toward them, and acted instantly by shoving her way between them and quickly shouted to her boss, "Jerry's right sir! I've seen the latest views of Hubble VII's deep space images of supermassive black holes, and those waves leaping out of black holes of that magnitude jettison out from the vortex so fast it's too unbelievable to believe sir!"

John just shook his head, and replied, "You know what I think? I think the both of you need to chill out now, because you're both starting to really tick me off!"

Jerry and Barbara quickly glanced at each other like he was nuts, and all Barbara did was shrug her shoulders to him.

Jerry swung back to him and yelled, "John! Under documented orders directly from General Vaughan himself, I am the lead appointed helmsman of this vessel, and I'm asking you to remove yourself from my post! Now would you please get up, and please let me do my job?"

John slowly shifted his look of anger up toward Jerry, and shouted, "Jerry, would you just please chill out and…"

Suddenly a loud warning signal alarm went off. John and Jerry quickly glanced at the alarm, and the indicator showed a massive erupting discharge rising out from the immeasurable vortex, that just now passed the black hole's event horizon and was heading upward at an unbelievable velocity.

John leaped while shouting, "All yours Jerry!" But in trying to get out of Jerry's way over to the copilot's seat so fast, he tripped over the copilot's chair, and while tumbling downward the side of his head slammed into the arm rest in the chair just behind the copilot's chair. He knocked himself out cold, and fell limp to the floor.

Jerry saw what had happened, but here was no time to spare to go to his aid. He jumped in his seat so fast, and while quickly strapping himself in, he shouted, "Barb! Copilots chair! Everyone else strap yourselves in now!"

Jerry was moving his hands across the instrument panel so fast it almost looked like a blurring haze. He saw that Barbara was having trouble with the straps, and rapidly shifted his swift moving hands over to her and strapped her in so quickly it was almost magical.

He then pointed down to a small group of switches, and shouted to her, "See those switches? Snap every one up as fast as you can!"

Barbara's cool quick action then took her over, and her moves also matched his, and within three seconds shouted back to him, "All switches up!"

Without looking to her, he then shouted, "Good girl! Now, see that line of buttons right above them? Push in every one of them, but do it from left to right only! Left to right only sweetheart!"

He then reached down to the main engine's two throttle levers, and slammed them forward as far as they could go, just in time to hear his copilot shout, "All bottoms pushed!"

The *Archimedes* quickly began to shake like never before, as the main engines began to instantly thunder their way to up full power.

Miranda shouted from her seat behind him, "Anything I can do?"

Jerry swiftly shouted back, "Just keep those straps as tight as you and Steve can make them."

Steven then shouted, "Where's Dr. Stanley?"

Jerry glanced to his left, and saw John still lying sprawled out on the floor, knocked out cold, not moving at all.

'Damn,' he thought, still moving his hands onto two separate dials.

He quickly shouted, "Barb, look at the rising discharge's proximity approach monitor! Over it are three buttons! Push in the left and middle button at the same time, and you'll see and countdown pop up! Tell me what those numbers are!"

She moved now with amazing speed, and instantly yelled back, "One hundred twenty eight seconds, one hundred twenty seven, one hundred twenty six…"

Jerry moved so unbelievably fast now by unsnapping all his belts while shouting to Barbara, "You have the con Barb! Be back in a few seconds!"

He then leaped out from his chair, and ran as fast as he could over to John's limp body, picked him up and slammed him in to the same chair that knocked him unconscious. Jerry snapped his chair's belts in under three seconds, then ran back and jumped back into his seat.

While snapping himself back in, he shouted to Barbara, "Request permission to take back the con!"

Barbara shouted, "Yes and way to go Jerry! You rock!"

"Don't thank me yet little sister! How much time left!"

"Ninety nine seconds!"

"Damn," he quietly said under his breath.

"Didn't like the sound that, Captain," Barbara quietly said back, "We're not going to make it?"

Jerry just grinned, and said back softly, "Are you kidding little sis? Of course we're going to make it! Don't you know that you and I are too pretty to die by being trashed by that throat of death? We're just going to have to use alternate plan B! How much time?"

With a short laugh, she quickly said, "Seventy nine seconds!"

"Okay," he snapped out while moving his hands again so fast again it was spooky, "Here it is, so listen very carefully. I have the ***Archimedes*** at full velocity trying to get us pass that oncoming gravity wave. Not going to happen, so here's the plan kiddo."

He then turned he navigation wheel hard right, while still talking, "I'm now turning us to a perpendicular angle, so when it smacks into us it will hit the stern first. Now, I just rerouted twenty percent of the main power to all the port and starboard thrusters."

Barbara interrupted, "Wait! The outer doors for those thrusters have to be opened!"

Jerry's grin never once faded, and said, "Thanks to my competent copilot, you already opened them, and every port and starboard thruster you've already brought online. Now, that reroute to all those thrusters is going to slow us down a little, but those thrusters you have ready are what's actually going to save us. Okay little sis, how much time left?"

"Forty seven seconds!"

"Good! That means it's starting to slow down a bit, and beginning to dissipate a little. See those two small levers there? That's the port and starboard thruster controls. When I tell you, put both hands on each one, and shove those puppies forward as hard as you can!"

"Can do Captain," she shouted back with joy, "Both hands in position now, awaiting your orders sir!"

"No, take your hands off of them and look and the countdown. I need you to tell me when we get to the ten second mark, and then grab them and wait for my signal to shove them both forward!"

"Twenty seven, twenty six…."

Jerry said, "Just tell me when we're at ten seconds baby girl!"

He then shouted, "Alright! Anyone know how to surf?"

Miranda shouted, "Not me, and Steven's fainted!"

Jerry shouted, "Me either! So let's hope we don't wipe out on our first lesson! Here it comes, so hang on!"

Barbara kept shifting her head up and down with each count, and slowly then raised her two hands toward the two levers, and then shouted, "Ten seconds!"

Jerry then jerked the navigation wheel all the way back, raising the ship's bow up slowly, while shifting his head with each mental count as she did, and then shouted, "Now Barb!"

The earth quaking thunder that came from the enormous blast of the port and starboard thrusters was so deafening, shaking the *Archimedes* like it was going to fall to pieces, until a massive powerful forward lunge hit them from behind, and the bow began to lower itself back down the toward a more level position.

Throughout the shimmering *Archimedes,* things could be heard breaking and shattering everywhere, but as the bow finally settled down back into its original position, all the tremendous shaking gradually began to fade back down.

The ship was still trembling somewhat, and Jerry said, "Okay, baby sister, pull both of those levers all the way back to their original position!"

When she did, the ship finally settled down again, and everything became perfectly smooth as if nothing ever happened.

Jerry quickly danced his fingers around to several more instruments, turned the main navigation wheel back to the left, and then locked it safely into place and dropped back into his chair with a few heavy sighs while quickly removing his straps.

He then jumped from his chair, and began to run around it as Barbara asked, "Are we out of it?"

Not looking back to her, he said back to Barbara, "We're good," but kept running over to where Miranda was sitting, kneeled quickly down to look her directly in the face, and asked in a frightened stuttering shout, "Are you okay?"

Unbuckling her safety belts, she replied, "I'm good. I think Steven's still out."

Barbara's shouting right next to them turned their attention toward her, "Steven! Steven! Wake up!"

She then slapped him across the face a couple of times, and he started to moan his way back to consciousness, and finally asked, "Barbara? Is that you?"

"Yes! Now wake up!"

He rose up opening his eyes, and said, "We're not dead? We're all still alive?"

"Yes," Miranda said to him, "But now your body is a long strand of spaghetti from being sucked through a black hole."

Jerry regained his self control, knowing the crisis wasn't completely over yet. He then looked to Miranda, and said, "Guess I better go check on the boss."

He turned and quickly walked over to where John was still out cold. He shook him a little, shouting his name a couple of times, but he remained unconscious.

Jerry unbuckled him, draped him over his shoulder, and headed for the infirmary with Miranda walking behind him.

In the hallway just outside the bridge, she said, "Hang on a minute Jerry."

She then lifted up John's eyelids, and then said, "Okay. Let's go."

Once there, he lowered John gently down into one of the infirmary's hospital beds, and Miranda began checking him out, since she was the appointed med tech of the crew.

Jerry just sat there watching her, to see the outcome of his exam.

After she finished, she sat down next to Jerry's worried face, and said, "It's only a mild concussion. He'll be okay after a couple of days rest."

"It's just that I don't know how to face him when he wakes up. John's not only my boss, but he's more my friend now."

Miranda then said, "You saved his life. You saved us all. I think he'll consider that when he comes around."

"Hope you're right. I sure think a lot of that man. He's the greatest boss I've ever had, and I hope it stays that way. But anytime he's ever been wrong, he's always the first to admit it."

"Why don't you go and get some rest," Miranda suggested, "I'll let you know how he's doing."

Jerry started for the door, and said, "Wish I could, but I need to perform a structural damage survey now, just to make sure if there's anything that needs to be fixed right away. Let me know how the boss is doing."

He then opened the door, but stopped when Miranda said, "Wait. Hang on a second."

She wiped her hands off, and then walked over to him, and said, "Listen. What you did was the most incredible act of bravery I've ever seen."

"Thanks Miranda," he replied with a shy smile, "I have moments."

"Like nothing I've ever seen," she said back, "And one more thing. Whatever the reason is that you haven't seen anyone in these past three years, if you ever want to talk about it, I'm always been known as a pretty good listener."

Jerry tried everything he could not to expose a full smile, but the best he could do looking directly into the most beautiful eyes he'd no doubt ever encountered, was a half of one. He just sighed as that half a grin gradually climbed up one side of his face, and stuttered with a major tone of shyness, as he softly replied, "I, um, just might take you up on it sometime."

Miranda pulled her head back a bit with a small look of confusion, and said, "Well, good luck with the damage survey. Hope things aren't too bad."

"I think we're pretty much okay, but with the ship taking a knockout punch like that, it's better to be safe than sorry. I'll let you know."

"Great. I'll get the boss back in shape. I'll let you know how he's coming along."

With a quick nod of acknowledgement, he turned and stepped out of the infirmary, and began heading back for the bridge. Just as he rounded one of the hallway corners, he met up with Barbara.

"Hey, Mr. hero! How's the boss?"

"Miranda says it looks like a mild concussion, but he should be alright. She said in a couple of days he should be fine and back in action."

"I was just now headed there to see, and also to give you this."

She reached out with both arms and hugged him hard, while saying, "Way to go, my new big brother! That was so incredibly awesome! I've never, ever seen anyone rock like that my whole life!"

"Well, I had one hell of a copilot at my side. You really rocked yourself, little sister."

"Thanks. It was an honor to help the man with not only the coolest head on his shoulders I've ever seen, but you've got more guts than anyone I've ever known."

"Well, with you at my side, how could I go wrong?"

"Thanks so much, but that's something I wanted to ask you about with no one else around. Why me? I mean when you seen what you knew you had to do to save us, why did you order me to the copilot's chair? Don't get me wrong, because I was so flattered, and still am, but I was just wondering. Why me?"

"That's an easy one. Think about it. Your little man passed out, which I'm sure we both know he was going to at some point, and could you imagine Miranda ever taking instructions from me?"

"I see your point. That was quick thinking too."

"But you've always had a good head on your shoulders, and a strong boldness about you. So I just had to do some quick math, and I mean real quick, and you got first prize little sis. So, how's Steve doing?"

"Got him back to his room, and made him take a chill pill. He'll be out cold for a while, but he'll be okay."

"Good to hear. Well, I've got to get going. The boat needs a good stem to stern check up."

"She couldn't be in better hands. See ya!"

CHAPTER 11

FOREVER'S EDGE

In the next few days that passed, John Stanley had made a full recovery, and now worshiped the ground Jerry Danielson walked on for saving his life, and everyone else's and the ship. He had already prepared the letter to General Vaughan recommending Jerry for the Congressional Medal Of Honor, that John knew he would no doubt get.

The ship was fully repaired, with the exception of every one of the outer doors that covered all the port and starboard thrust exhaust chambers that ran down each side of the ship from bow to stern were now gone, blown instantly off their opened positions when the surge from the supermassive black hole struck the ship from the rear side.

They had now traveled thirteen point six nine billion light years from earth, and were well under an hour away from passing the last known quasar, ready to make the next historical discovery in the history of their civilization's long existence.

But something was very wrong. All sitting in the ship's main conference room discussing the next move, Barbara had everyone's attention.

"Every scientist as far back as the twentieth century has always claimed that if you travel thirteen point seven billion light years," Barbara explained, "You will in fact have gone back thirteen point seven billion years in time, and actually witness the birth of our universe because you'll then be a spectator of this so called, 'Big Bang' explosion.

"Steven and I have been up all night sending the ship's long range scanners in every direction, and every reading that we received back returned no images at all. Nothing. Only the void of empty space ahead, and that's it. So where is this magnificent, 'Big Bang,' explosion?"

"I've never believed in it either," Jerry said, "Because if Hubble's eyes can image pieces parts at the end of the universe, and claim they're the first objects to exist as a result of this Big Bang detonation, why can't Hubble ever see some kind of explosion? Or even signs of one?"

The conference room went dead silent.

"Alright everyone," John said, "It's a little under ten minutes left until we reach the thirteen point seven billion light year mark. Let's follow this through so we can say we went all the way. Grab your coffee and let's head for the bridge. That will make it even more official."

Walking side by side in the corridor toward the main bridge, Jerry asked John, "So how are you feeling now? Pain ever lighten up any?"

"Jerry," John said mildly smiling, "You don't have to keep asking me that. I'm fine, okay? You saved my life, and I'm getting tired of hearing your guilt eat at you because of something that wasn't even your fault. I fell out of my own clumsiness. Now would you please stop it?"

"I wouldn't keep asking if you weren't my good friend boss."

"Thanks Jerry. For someone that's in a situation that doesn't know how to handle it, you sure knew how to handle that one. By the way, I haven't heard Miranda say one ugly comment about you since I woke up. Any luck with her?"

Jerry smiled from ear to ear, and said, "No, but she offered to be my listener if I ever want to talk about why I haven't been with anyone in over three years. Can you believe that?"

John smiled back, and replied, "That's probably the most ironic thing I've ever heard."

In the main bridge, John was in the front pacing back and forth as he looked at his watch counting down.

"Three, two, one. Okay people," John began, "Here we are, and I for one am so disappointed, because the only thing ahead is clear empty space. Suggestions please, for our next move?"

The other four just looked at each other, wondering who was going to be the first one to suggest anything.

John then said, "C'mon, guys! We have the ability to go wherever we want, and none of you can think of somewhere or something you'd like to see next? Steven? Miranda? Jerry?"

Jerry replied, "Don't look at me guys. I'm just the chauffer here. Just point, and I'll get us there."

Steven's always well mannered character was in full force, and politely raised his waving hand.

John always acknowledged his respectful quality, and said, "Yes, Dr. Meyers?"

"Thank you sir. I suggest we just keep going."

John turned around to the view window, then looked back to him along with everyone else somewhat puzzled.

John then asked, "Keep going where?"

Steven replied, "Remain on our same course, and just keep going, sir."

"But Steven," John smiled as he replied, "There's nothing out there."

Steven really had to work up his courage for a few moments to tell his supervisor something that he thought he would never say to him, "I, um, disagree. Sorry. I mean I disagree, sir."

John thought for a moment, and then said back seriously, "I don't mean to sound sarcastic at all, Steven, but do you really think there's something out there?"

With all eyes on Steven, he cleared his throat, and then explained, "Every since I was a child growing up on Earth living in Ohio, my Dad and I would go outside after dinner and wait for the stars to come out. Once they did, my Dad would teach me something new every night.

"For instance stars twinkled and planets didn't, why the moon went through different cycles of how much light it reflected due to how much of an angle it was aimed toward the Sun, things like that.

"But the one thing that neither he nor anyone else could ever explain to me is why the area behind all these celestial bodies were solid black. The only response I've ever got was because it was the blackness of empty space, which to me was never an answer. There had to be another answer as to why everything behind all the bodies in space in our universe was black. To this day I could never find a concrete answer, but I would still like to know, and believe if we keep going we'll eventually find out."

The whole room was completely silent. John continued to pace back and forth with a look of serious curiosity about him.

John then slid himself back on a little table right next to the control panel where the pilot sat, and said, "You know, that's one interesting question that really never occurred to me. I mean blackness indicates an absence of color, yet I agree. Why is it black? Opinions anyone?"

"Some would say it's because the universe is infinite," Barbara said, "But that still doesn't answer Steven's question. I mean there must be some kind of, um, what's the word I looking for? Not really a wall, but more like a, um, let's see…"

"Threshold," Jerry suggested.

"Bingo, big brother," Barbara replied, "Like perhaps into another type of dimension, or something like one, that we could never even begin to comprehend."

"That's an interesting point," John responded, "How about you Miranda?"

"Never really gave it much thought," she said now in a serious tone, "But if I had to guess, I would probably say that there are other universes, maybe just like ours. However, they're just so, so far away from our place in this certain realm of wherever we are, that there's no possible way we could ever detect them from here. So, whatever you guys decide is fine with me."

"Fair enough," John replied, "And Jerry? Just to make sure, if Barbara, Steven and I decide to just keep going, you're good with that too?"

Jerry responded, "I'm with, I mean like Miranda's said. Whatever you guys decide is fine with me."

With that same puzzled look Miranda gave him that night he responded to her offer about hearing the reason why he'd hadn't been with anyone in over three years, John swung her attention back his way, as he said, "Great. Now Barbara, I'm all for it, and it must be a unanimous decision. So it's your call. What do you want to do?"

Barbara had to turn to the side for a moment to try and hide her grin, but Miranda shouted, "I see that big smile girl! C'mon, Barb, let's hear it!"

She turned back and started laughing a little, and then said, "Of course I'm for it, you know why Mir?"

She then leaned her head over and rested it on Steven's shoulder, and then said in a somewhat sweet tone, "Because I'm always with Steven."

Before the room erupted in laughter again, John quickly shouted, "Alright, let's please keep it sane people. Okay, into the black void we go. Jerry, make sure we stay on that same straight line course we've been on that we established between the Milky Way and Andromeda, so we'll have no problem finding our way home."

Jerry climbed into the pilot's seat, adjusted a few controls, and then confirmed, "Course lock unchanged boss. How fast? Maintain full speed as we're doing now?"

He just shrugged his shoulders, and then replied, "Why not? Maintain full power."

The *Archimedes* then soared its way into the completely unknown black void of empty space, and within a mere sixty seconds was almost a half of a million light years away from their own universe.

CHAPTER 12

OCEAN OF DARKNESS

A few months had now passed since they blasted away from their universe into this blank, dark void that now engulfed the **Archimedes.**

Jerry stood in front of the main control panel, leaning his back against it with his arms folded and letting his mind drift aimlessly into the void, wondering just how much further they had to go before realizing there was nothing out here.

Hearing the rear door open, he turned and saw Miranda walking in. Shutting the door, she walked toward the front saying, "Hey. Checking out the impressive view?"

"Oh yeah," smiling as he spoke, "Pretty charming, eh?"

"Really," she said, and kept walking toward the front viewing window, "Awesomely strange. Steven's sure we're going to find something though. Any word as to how much further we're going to go before we throw in the towel and head back?"

"No," he said back, with a mild tone of frustration, "I mean if you ask me, this is far enough."

"Ditto on that," she answered with that same touch, "Seriously, what have we found out here? Nothing. Except for an occasional rouge space rock, or a stray pocket of gas zipping by us every once in a while, and that whatever it was Barbara picked up on the long range scanner weeks ago."

"Six weeks ago," Jerry replied, "I keep wondering about whatever that was. But it shot by us so fast, the scanner didn't have time to get any reading at all. By the time she got John and I in here, it was so far behind us the scanners didn't register anything. Only a split second recording. Other than that, its been really uneventful out here, wherever here is, and I'm ready to head back home."

"Ditto again," Miranda agreed, "We really need to bring this up when John's up and around. I'm ready to head back as well. Just for the record, how long have we been cruising through this ocean of darkness?"

"Good question," He said as he swung around the operations control panel and sat in the pilot's seat, and then verified, "Exactly three months today."

Miranda walked over and propped her elbows up on the opposite side of the control panel, and leaned down enough to look him directly in the eyes, and calmly said with a small cheerful tone, "Sounds like a familiar, 'moment,' I believe is the phrase you always use. That's pretty much how long it's been since I offered to listen to your reason as to why you haven't seen anyone in over three years. I believe your answer was, 'Just might take you up on it sometime.' Well, for what it's worth, the offer still stands."

Jerry was not just a little over a foot away her, he was looking her directly in the face, rapidly turning into a pathetic pile of putty. She also had on something that he rarely ever saw, and on those extremely few occasions it was always from a distance, mostly at formal gatherings were he knew she always made it a point to stay as far away from him as she possibly could.

It was makeup. Not too much, but a small brief touch of eyeliner, a little eye shadow, and most of all few swipes of lipstick.

He was so awestruck with the immeasurable feelings of love he felt for her, but combined with just how beautiful of a woman she truly was, captivated him into a completely hypnotic trance that all he could do is stare straight back at her.

"Jerry," she finally asked, "Are you alright?"

He somehow activated his awareness by first looking to the side and saying, "Yeah! I'm good. It's just that's a hard one for me to, um, talk about."

"Oh, okay. I understand. Well, sorry, because if it's that intense, I'm leave you alone about it."

She then turned and started walking toward the exit.

Jerry knew this was it. Enough is enough, win or lose. Just before she got to the rear door, in a sudden state of panic he jumped up and shouted as he turned backwards to her, "Miranda! Wait! Please?"

She stopped while turning back around, and with a friendly laugh replied, "Of course, Jerry! What's got you so wired that you had to almost yell?"

"Sorry," he softly snapped back, and then took a quick deep breath, and then asked, "Would you mind taking a seat? I have a feeling you're going to need to sit down when you hear this."

She swiftly exclaimed, "What? Oh, this must be good! No, not at all," she said with an ongoing giggle of surprise, and walked back and sat in the chair just behind the pilot's chair.

Before falling back into that trance again while now looking down at her, he quickly began, "Okay. A little over three years ago, I wasn't so nice to women. I was chased so much and thought how could life ever get any better than this. That is, until something I never thought possible happened.

"I met this woman, and fell so in love with her from the moment we were introduced, and knew then, I mean at that very second, that this is what it was supposed to feel like. She was so incredibly beautiful, I wondered and still do, how in the Holy name of God could He have made a woman so gorgeous that she actually looks like a goddess."

"I can't believe what I'm hearing," Miranda said with surprise, "You, head over heels for a woman? Well, what happened when you told her?"

Jerry swallowed hard, and answered, "She doesn't know. I mean I've always been chased, okay? And to this day I don't know how to go after woman. But one thing's for certain. If I can't have her, I do not ever want to be with another, and it's been over three years now."

"I still cannot believe this," Miranda said back now with a touch of sadness, "I mean I do believe what you're saying, but Jerry, I've lost count on how many news flashes that would announce, 'Again, Dr. Jerry Danielson wins the, 'Sexist Man Alive Award.' But you're telling me you're going to wait and see if this woman has a change of heart, when she doesn't even have the slightest clue that you feel that way about her?"

"Is it really so wrong of me not to know what to do?"

"Jerry, you are one remarkable man. Stuck in a paradox like this, having so much faith that somehow, someway it will work out. But if you ask me, you should just tell her. She really that pretty?"

He kneeled leisurely down to her, and said, "Miranda, you have no idea."

They just stared at each other, as she was now really starting to see just how incredibly handsome he was, and what began to spark a real flame in her heart now is that she knew it was coming from another honest sincere heart.

She broke the comfortable silence with, "Well, when we get back home, you've got to tell her. You have to tell her, because if you don't, I'm going to be real disappointed in you. Do you hear what I'm telling you? Do you really hear what I'm telling you, Dr. Jerry Danielson?"

He sadly shook his head yes, and said, "Yes, I do. You want to know why? Because Miranda, I don't have any choice."

With a sadness trailing in her voice, she said, "Well, I really hope things work out for you Jerry. I really do, and when we see the stars of home getting closer…"

Jerry thought 'well, here goes', and then interrupted, "Miranda, please listen to me. It's not the stars we'll see as we get closer to home that will make me happy…"

"Speak for yourself," John's voice echoed throughout the room to their surprise, as he staggered through the rear entrance door still half asleep, and added, "I think three months is more than a fair enough shot at this one, not to mention I'm starting to get the creeps. I've awakened Barbara and Steven to meet us here, and I told them what we're doing. Turn us around Jerry. Take us home."

Jerry and Miranda looked back to each other, knowing he was only a few words away from telling her, but what they both just heard turned their attention now in a completely different direction.

Jerry rose from his feet, and acknowledged, "My pleasure boss," and quickly turned around and jumped into his pilot's seat. His hands then began to dance around the ship's navigational controls.

Barbara came into the bridge with Steven yawning behind her, saying through her voice of exhaustion, "Steven are were just talking about how both of us would like to see some stars and galaxies in our viewports when you called us Dr. Stanley."

"Hang on a second boss," Jerry said while looking into the small viewing scope that protruded up from the control panel's left side, "I've got something here."

Suddenly the long range proximity sensor alarm began to give off its pulsating noise and began flashing red.

"What do you mean by something? What is it," John asked as he and everyone else began to quickly encircle him.

Jerry jerked his head back from the viewing scope with a look of shock, and then quickly lunged his head forward again to confirm to himself what he just saw. He quickly tapped one of the control panel buttons, so the image that he was looking at could also be seen on the huge monitor over the main viewing window.

He then said so softly shocked, "Can you believe what we're looking at?"

Gradually coming into view started as a small shimmering light. The closer they got to it, it began to take little familiar shapes, so Jerry zoomed in on it as far as the ship's telescope would allow.

The whole team now gasp along with him now; magnificent galaxies and bright glazing quasars. Beyond them they could see expanding gleams of light.

"Well," Barbara said, "Steven was right. Correct me if I'm wrong, but it sure looks to me like another universe."

Steven then said, "Yes, it does. But there's something not right. Jerry, can you spilt the view so that we can see what our universe looked like when we pulled away from it?"

Jerry adjusted a few controls, and what they saw now was even more amazing. The two images were completely identical, except the one on the left side was coming toward them, and the other was moving away on the right.

John walked closer to the monitor, and asked, "Please tell me the one on the left is the live view and the one for headed for."

"They're identical," Miranda said straining, so captivated like the rest of her team.

With a similar tone of alarm, Barbara quickly asked, "Then what happened? Have we somehow been going in one big circle these last three months?"

Everyone just stared at the two images, still trying to accept it.

John began walking a little closer to the unbelievable image on the monitor, and mumbled very slowly, "What the hell?"

He then spun around quickly asking, "Jerry, do we still have a secure lock on the return course?"

Jerry looked down, adjusted a few controls, and then slowly looked back up to John, and answered, "Navigation controls confirms return course lock verified boss, but…"

John instantly shouted, "But what?'

Jerry swiftly swallowed, and then replied, "But, they're aiming in the opposite direction! The direction we're coming from!"

John along with the rest of the crew, all with the same astounded faces quickly turned back to look at the approaching universe.

"Well," John asked, "If we have somehow gone around in one big cosmic circle, I don't see turning around now and traveling another three months just to see if we're right! Anybody else?"

Everyone all shrugged their shoulders while turning to look at each other at the same, and one by one agreed.

John then looked to Jerry, and said, "Take us in on the same course we came out on, and let's just hope in three weeks we're home."

CHAPTER 13

HOME FINALE

The welcome home ceremony was nothing less than splendidly exquisite.

Upon the splendid dining table was almost every type of superbly cooked food.

Sitting front and center was John, with Barbara on one side and Steven on the other, then Jerry on Steven's right side and Miranda on Barbara's left side.

Barry Harper was directly across from John, with General Vaughan and a few other White House guests spread out on each side of Barry and Vaughan, all cheerfully enjoying this memorable moment for their return.

Somewhat lit by the priceless rare red wine, Barry couldn't stop talking, "We really thought you had reached the Big Bang and all of you were instantaneously vaporized! I mean four and a half months you guys zipped away from here, four and a half months in the fastest vessel, better yet the first 'starship' all the NASA folks and affiliates could ever construct!"

He then looked quickly at Barbara and Steven, and added, "Of course that didn't make the ***Archimedes*** the supremely remarkable ship it is, without the astounding discovery from you, Dr. McKenzie, and you, Dr. Meyers!"

General Vaughan took it from there, "Supreme, remarkable, astounding, or any word just does not give it the proper justice it truly is. In fact, there are no words that can honestly represent her. None. That incredible ship John is what we all thought, and I'm sure yourselves included, of being centuries ahead of its time! Centuries John!"

Smiling proudly as his fist pounded a few times on the table for silence, John raised his glass of wine up, and announced, "Here's to Barbara and Steven! A match made in heaven! Yes, general! I'm quite sure I speak for the rest of our team, countless centuries ahead of its time. Even more, if I had to guess! Way more! Guys, care to add anything to that?"

Jerry stood up and shouted with his glass raised, "Being that it happened in our lifetime is nothing less than an act of providence! And we are the first ones to experience the unbelievable velocity Barb and Steve came up with! Honestly, part of me still doesn't fully believe it. But what to hell everyone! Again, here's to my little sister Barb, and my main man Steve!"

Everyone cheered and took in a heavy swallow, as Jerry sat back down with his wine glass fully extended now straight over his head.

Barry then raised his glass shouting, "And to live to tell about it!"

Everyone cheered taking in another full gulp. Everyone except John and his team. He looked around the dining room, realizing the most of the people wasn't sure by now what they were actually cheering about.

He then shifted his barely smiling face to his team on each side of him, who were all smiling about the same level as John. They raised their glasses to conform with the rest of the small crowd, but took in only a small sip. Just as they sat their glasses back down with their wrists now resting on the table, they carefully raised their hands up to John, signifying to him that they didn't know how to tell them.

With a fake gesture of joy, John then shouted, "Hang on a second people! Can we hold it down for a minute? This I'm quite sure all of you need to hear, a, well, fantastic discovery if you wish to call it. But please everyone, this is going to knock you for a loop!"

The crowd quickly calmed down enough to almost a silent moment, with all eyes now upon him; especially General Vaughan and Barry Harper's.

"Thank you," John said quickly with another counterfeit act of courtesy, "It's what happened after we completed our first assignment, which all of you know what our objective was."

Barry still laughing intervened, "Of course, to go all the way to the edge, to start with settling all the rumors these past few centuries about the Big Bang theory. Every since we made contact with you returning, you've been dancing around that question at every attempt I've made.

"So before I squeeze it out of you, now would you tell me," he hesitated for a quick second with a small burp, and continued with a bit of a slur now in his words, "Sorry John! Now, what was it? Oh, what I meant to say is would you please tell me, I mean us, what it was that you guys saw when you first got to the edge?"

"It's what we didn't see old buddy," John reluctantly replied, "Nothing. Nothing of any significance. Beyond the last quasar, which is actually thirteen point six eight nine seven billion light years from earth, all we saw was empty space ahead. No Big Bang, no explosion of any kind. I mean not even a stork delivering a baby with the true secret of just how our universe began."

Then room went dead silent. No reactions of any type. Just a table of the highest ranking VIP's, all with frozen faces.

General Vaughan then broke the silence, "Wait a minute. No Big Bang, or anything familiar to it?"

"Nothing General. Nothing but the big black void of empty space was ahead of us. We then all voted to just keep going, and to remain precisely on the straight line course we established between the Milky Way and Andromeda. Here, take a look for yourselves."

He then turned around and asked the security officers if one of them would mind shutting off all the lights. A few seconds later the room was shrouded in complete darkness, and John then clicked a remote control, and a viewing screen instantly rolled down behind Barry and the rest of the guests.

A quick second after the bottom of the screen was all the way down, everyone on Barry's side had already turned around. A display of a very small group of galaxies and quasars appeared on the screen, drifting by until it passed the very last quasar.

John then paused it.

He went on, "This, gentlemen, is exactly thirteen point seven billion light years from earth. Now, what do you see ahead? Nothing but empty space. The identical view of what the Hubble VII showed since she went on line."

The entire room was dead silent, and John continued, "Now, anyone care to speculate what this means?"

Somewhere toward the end of the left line of Barry's guests, one of the White House reps' voices was heard through the darkness, "That time is constant. No past, no future. I've always believed that, and now all of my disagreeing colleagues will now have permanent concrete proof. Oh, I cannot wait to see the looks on their faces!"

John said, "True you are sir. But I'm afraid it gets a lot better than this."

"It should," Barry remarked, "It took you guys three weeks just to get to the edge, and it should have taken you three more weeks to get back. But John, four and a half months? Just how far did you people go into the void of empty space?"

"Three months exactly," John replied, "Ironically, the day I walked into the bridge and ordered Jerry to turn us around, just as he was adjusting the controls to do so, Jerry then said the long range censors just picked something up. This is what we scanned that we were heading for."

After showing the video of the universe returning into view the exact same way they left it, General Vaughan advised all the VIP's present not to repeat what John just revealed to them, until a thorough investigation by he and his staff was performed. He himself would contact the president, and only in person.

After the dinner, General Vaughan asked John if he could speak to he and his staff privately, without Barry Harper.

Once they were all behind locked guarded doors, the general said, "I apologize John to you and your historical crew here for holding all of you up since you just returned and I'm sure you have personal things to attend, so I'll be very brief.

"I just want to hear this once more from you just so I have this perfectly clear, without the annoying voice of Barry Harper's alcohol interference. Okay. You guys pulled away from here and it took you three weeks to get to the edge of our universe, then three months into the black void of space, and then three weeks from the edge of the universe to get here.

"Correct," John answered.

General Vaughan's eyes swept across each face, and then continued, "Your course never deviated at anytime?"

"Not so much as an inch," John replied but now looking at Jerry, who was shaking his head yes.

The general rose up and began walking slowly around the conference table, and then asked, "Any ideas or theories as to how you somehow got back here after traveling in a straight line the entire four and a half months?"

John replied, "We've checked every instrument from stem to stern, and everything checked out fine. The only two conclusions that were established in that instrument check were, number one, space outside this universe somehow curves back around on itself, and number two, we just plain don't know."

General Vaughan said, "Well, something fantastic has happened here, or out there. Okay, I'll have my people look in on it."

"Unless," Jerry said, "It had something to do with my cough."

"What," John asked him, and the general stopped pacing.

Jerry replied, "You know sir, my cough. Every day I sat at the wheel, I drank five bottles of prescription cough medicine a day, but I'm sure that had nothing to do with it."

The entire room burst out laughing, especially General Vaughan, who walked around the table opposite from him and leaned down on it to face Jerry directly, and said with a sense of powerful pride, "Jerry, it's times like this I always think of my best friend who I grew up with, and you are so a mirror image of him because he had your exact magical wit.

"He too was also a hero many times over, because during any crisis he always knew exactly what to do. Thank you for that little reminder, and I still miss him very, very much Jerry. I just wanted you to know that."

Jerry then looked down with a face of sadness, and with a little choke in his voice, he sadly said, "I miss him very much too, general. Thank you."

"It wasn't my intention to get you down Jerry," General Vaughan said, "It's just I've really never made that clear to you, because it's so very hard for me to talk about.

"But your father, Dirk Danielson was the finest man I've ever known. A true hero, just like his son. Congress and the president approved your Congressional Medal of Honor, and the president is so happy that he gets the honor of awarding it to Dirk's son, as he did to Dirk almost four years ago. I'll let you know when the award ceremony will be."

"Thank you general," Jerry said a little more cheerfully.

General Vaughan leaned back up, and said as he started walking toward the door, "That's it people. Thank you all.

I'll let you know what my people come up with about how the ship went in a big circle John."

He placed his hand on the door to open it, but stopped and turned around, and asked, "In the meantime, is there anything else I can do for you people?"

Jerry answered with his same witty cheer, "Yes, just one thing general. Your people that's going to "look in on it," ask them to try and not blow up any more moons."

Everyone mildly laughed, while General Vaughan smiled again with so much pride, and said, "John, you know what scares me the most whenever Jerry and I talk? He looks exactly like his father, along with that same voice. Let me know Jerry when you can get a little free time. I'd like to take you out for a few beers, and tell you some pretty tall tales about your dad that I bet you don't know about."

"Thank you general," Jerry replied, "I'd like that."

Once the general walked out the door, John just sat and stared at Jerry.

Miranda, Barbara and Steven rose from the table, and Miranda asked, "Okay. Are we all done?"

John remained motionless, still staring at Jerry as he also slowly rose up.

Everyone then stopped, since John still didn't dismiss them yet, and still had that stone cold glare on Jerry.

Jerry then asked, "Well? Are we done?"

John finally spoke, and said with such a serious tone, "I can't believe you."

"Can't believe what," Jerry asked.

John then rose up to face him directly, and asked, "Why haven't you ever told me that you're actually Dirk Danielson's son?"

Jerry glanced down, and then back up, and replied firmly, "Because I miss him very much, and it's a little hard for me to talk about. Now can we leave it at that?"

John just nodded and turned to walk out.

Barbara's voice came from behind Jerry, "That's so awesome Jerry."

"Very awesome," Miranda added, "He's a legend Jerry. I'm so sorry he's gone."

Jerry stopped, and turned around to say with a small smile, "Thank you Miranda. Coming from you, that really means a lot to me."

He then turned and walked out the door. Miranda stood stiff in her tracks, a little shocked from what she just heard.

Barbara led Steven around her, holding his hand. She stopped and turned around, lifting her finger up to Miranda, and said, "I keep telling you Mir, but you're never going to believe it."

Miranda was grinning, and said, "Wow. Sure didn't expect that."

Barbara replied, "That's exactly why you'll never believe me."

Later that evening, John and Barry were sitting in the station's observation deck, taking in the marvelous view of Neptune's spectacular bluish white haze. Miranda took a shuttle to Earth to check on things around her beautiful home in southern Paris, and brought Barbara and Steven along.

Barbara had always wanted to see Miranda's extravagant home from all the videos Miranda had shown her, and Steven went along because it was getting to where he couldn't go more than an hour now without seeing Barbara.

Jerry was giving the ***Archimedes*** a brief check up along with a couple of Barry's technicians, and then took her out for a another test run when they decided where they were going next.

Laughing along with Barry about the dinner they just came from, John's com button went off, and also a little red flashing light on his button indicated an urgent private call from Jerry.

Not even saying his name, John answered, "Hey! What's going on?"

"Bossman," Jerry quickly shot back with the sound of immortal terror, "Are you alone?"

"No, but I'm kind of tired and I'm going to bed in a few minutes."

"Boss listen," he said with that same tone of shock, "There's something you've got to see, and I mean right now! I'm in the ship rounding Neptune, and will be at the boarding platform in five minutes. You've got to see this, and I mean right now!"

John thought of a quick acknowledgement, "Yeah, I know, and I'll always be there for you, and anyone else on my team. You know that."

"Great," Jerry verified, "See you now in about four minutes!"

"Okay. We'll talk more about it in the morning. Bye."

John told Barry he was quite exhausted, and going to hit the sack.

Barry then asked, "Sure John. What was that all about?"

He just quickly sighed, and replied, "One of my staff needs a pep talk for someone they're trying to get back with now that we've returned. Goodnight Barry, and thanks again for all the laughs and that fantastic dinner. I'll see you in the morning."

"No charge John," Barry smiled with pride, watching John head for the elevators, "See you tomorrow."

CHAPTER 14

RENDEZVOUS WITH REALITY

The beautiful regions of southern France were absolutely breathtaking. Barbara was so mesmerized watching the panoramic views of the rolling hillsides so smothered with enriched green pastures and forests below.

"Look," she shouted to Miranda and Steven, "We're getting closer to Paris! I can see the top of the Eiffel Tower way off in the distance!"

"We'll be landing in a few minutes," Miranda proudly said, "There's a landing platform about two miles from my house. That's where I keep my transport shuttle parked."

"Oh Mir, I can't wait to see your beautiful home! Right Steven?"

Steven was preoccupied on the ride with whispering phone calls, and as he clicked off his com button, he turned back around with a look of utter shock, and asked, "I'm sorry. What, um, were you saying?"

Barbara asked with a touch of confusion, "What's wrong baby?"

That caught Miranda's attention, and as she turned back to look at Steven, she saw that he was just shaking his head with a glare of mass confusion more as each moment passed.

Miranda and Barbara glanced at each other, and then Miranda asked, "Steven, what's up guy?"

He sighed once, and said, "It's just that, well, I've been having this really peculiar feeling every since we landed back on the Hubble Station that something wasn't right."

Barbara remarked, "Guess that explains why you've been so strangely quiet, I mean much more than you usually are."

"I'm sorry Barbara," he sincerely said, "But now I have that feeling even more now. I've been talking to my sister Penny, and the whole time she kept apologizing because she had no way of getting word to me because we were on our mission two months ago when it happened."

Barbara quickly asked in a sympathetic voice, "Ah, baby, what happened?"

He gradually raised his head up to her with that same look of such conflicting confusion, and slowly replied, "My mother died two months ago."

Barbara shouted as she exclaimed, "What? You're kidding me right?"

Miranda jerked her look at Barbara, and said in an angry low tone, "Barbara! It's his mother for Christ's sake! Ah, Steven, I'm so sorry! Come here!"

She sat her carryall bag down quickly, and just when she reached out to hug him, Barbara stepped between them looking Miranda directly in the eyes, and with the same look of shock Steven still had, she said in a strong whisper, "Mir! His mother died two years ago!"

"What," she repeated Barbara's first response, "You're joking! What did you tell your sister Steven?"

"Nothing," he said, "I was so shocked, I just went along with what she was saying. Besides, with Penny you can't ever get in a word edgewise."

Now all three of them had the same look of shock, when the pilot announced, "We're landing now Dr. Young. Please take your seats."

Walking across the parking bay for her personal transport shuttle, Miranda kept repeating, "Somebody just got their wires crossed. That's got to be it."

Every time she did, Steven would answer, "I don't know. I don't see how."

Soaring slowly over the exquisite landscapes below in her shuttle, Miranda concluded, "We'll make a few calls from my house and get this whole thing straightened out Steven, which we're coming up to now. There."

Barbara looked to see where she was pointing, and exclaimed, "Oh my God, Mir! Just like your videos! That's absolutely magnificent!"

Steven wasn't even looking as she slowly circled the extraordinary three story home so they could get a full view of the outside, and then reached down on the control panel and clicked one of the switches upward.

Two doors on the houses' reverse side opened up into a small parking area, and she drifted the shuttle in on auto pilot, which made a perfect delicate landing.

Letting Barbara and Steven out first, Miranda grabbed her carryall bag, and said, "C'mon, right this way! Hope the cleaning service has been coming as scheduled."

Miranda then opened the bay door and stepped into the house, and then stood sideways to let Barbara and Steven walk in ahead of her.

They all walked down a small hallway and in through another small passage, and as they walked into a huge beautiful room, Miranda announced, "And this, is my living room!"

Suddenly something grabbed all of their attention.

The sound of a barking dog getting closer every second, and before she realized it, a quick running beautiful little Pomeranian ran in where they all stood. It stopped about five feet away from Miranda, but continued barking fiercely at the three of them.

"What the hell is that thing…"

Her attention then jerked her look up behind the yapping little dog, when a woman dressed like a maid came walking quickly into the room.

"Oh, Ms. Miranda," the woman said, "I had no idea you were coming. I knew you were back, but usually you call me to let me know when you're on your way home. It's so good to see you!"

She then stepped up and gave Miranda a brief hug, but while looking over her shoulder Miranda glanced at Barbara and Steven with her eyes widened in astonishing alarm.

The woman then turned around and shouted at the wildly barking dog, "Mojo! What's the matter with you! You haven't seen mama in almost six months now! Now hush it up boy!"

Miranda quickly said. "We've all been through some serious decontamination procedures, and they said it will remove all of our body odors for a few weeks. That's all it is."

"Oh," the lady shouted back over the dog's vicious barking, "Well, as your maid I just wish I had known you were coming home now Ms. Miranda. Mojo! Hush it up now boy!"

"Well," Miranda yelled back to her, "That's another thing. I'm not really back yet. We just stopped by so I could grab a couple of file discs. I can't stay, so when I return I'll call you and let you know I'm on my way. Gotta get going now! See Ya!"

She quickly turned to Barbara and Steven with those same widened eyes, and motioned toward the hallway, but Miranda turned back long enough to shout over the dog's barking, "Oh, I'm sorry! This is Barbara and Steven!"

The lady shouted back, "Oh, I've seen them on the newscasts plenty of times with you! Okay, bye now!"

The three started swiftly moving through the hallway that led to the parking bay. Once the entry door was shut, Barbara quickly asked, "What's going on?"

All rapidly strapping themselves in as they got back in the shuttle, she said with a strong whisper, "Let me get us out of here first!"

The little shuttle then rose from ground while quickly spinning around to face the two separating bay doors. Once fully opened, they shot out and up so fast it pinned the three of them to the back of her seats.

Barbara shouted, "Mir, take it easy! What's going on?"

Through her heavy breathing, she said with a frightened, almost hysterical tone, "I don't have a maid! Never did, much less a little barking monster!"

She slapped her com button quickly, and then turned the little knob that signaled an outgoing private call, and yelled, "John, Jerry! Come in! Come in!"

"John here Miranda," quickly jumped out from her button, "Where are you?"

"I'm with Barb and Steven heading away from my house, heading directly for the space shuttle platform, getting ready to land now! Sir, you're not going to believe this…"

John instantly interrupted, "That we're not back in Kansas? I know! Jerry and I are aboard the ***Archimedes***, and he's changing course now coming in your direction. Do not land at your space shuttle platform! Just head up into the atmosphere as fast as you can and we'll track you by your com button and pick up you guys in a few minutes!"

Miranda raised the nose of her little shuttle straight up, switched on the anti-acceleration device and kicked the main thrusters almost to full, and shouted, "Heading straight up now sir!"

"Roger that," John replied, "We're almost there now, and Jerry has a good lock on your com tag. Opening up the outer shuttle bay doors now. See you in less than a minute!"

Slamming open the main bridge door, the three of them rushed straight up to where John and Jerry were sitting at the main operations control.

"Sir," Miranda exclaimed, "I brought Barb and Steven to show them my house, and there was a maid there! A maid, and a little Pomeranian that wanted to tear my head off!"

"Miranda, guys," John shockingly said back, "Wait till you see this one! Show 'um Jerry."

The ***Archimedes*** slid down and over the northern surface of Uranus, as Jerry said, "Oh, you guys are going to love this one!"

Approaching them rapidly as it filled almost the entire viewing window was one of Uranus's moons, a beautiful full circular disc.

John then asked, "Anybody care to guess which one of Uranus's moons that is?"

Steven said in an astonishing whisper, "Oberon? That's Oberon!"

Jerry replied in his usual voice of humor, "Give the man a cigar!"

Steven quickly replied, "But how?"

Barbara answered, "The exact same way your mother died only three months ago, when she actually died two years ago. The same way Miranda now has a maid and a dog, and now the reason we're all looking at Oberon. We're not home, guys. This is some kind of parallel universe, parallel to ours."

"Bulls eye Barb," Jerry said, "Now check out what the scanners indicate where the moon Titania used to be."

Steven looked down at the image indicator, and said, "Nothing but ionized dust and debris fragments. So our people blew up Oberon, and they blew up poor Titania?"

Jerry then shouted, "Give the man two cigars! Now, can I please get us the hell out of here boss?"

John smiled as he quickly said, "Only as fast as you can Jerry, before they start wondering just what the hell is going on!"

Jerry instantly pointed the ship's nose straight up, and blasted their way above the parallel version of the Milky

Way galaxy, and then adjusted the course precisely on the direct home line course they set between the Milky Way and Andromeda in their universe.

He then engaged the main thrusters to maximum, now watching the approach of the parallel galaxy of Andromeda begin to drift under them.

CHAPTER 15

THE MIRACLE OF FATES

Three weeks passed, and Miranda and Jerry stood leaning on the front cushioned rails that were placed in front of the large viewing window up on the observation deck. They had just cleared the last quasar of the astounding equivalent universe that was almost a mirror image to their own.

"Oh, God, am I ever so glad we're out of there," Miranda said, "No chance of our course being changed for some reason?"

"None," Jerry replied with a tone of sadness, "That homeward lock will get us back to where we belong. Three months, and three weeks to go."

Miranda, who had really fixed herself up when Jerry asked her if she could meet him up on the observation deck, said, "Can't wait to be Barb's maid of honor for her and Steven's wedding next week. And you too should be proud, being Steven's best man!"

"Well," Jerry replied in that same voice of sadness, "He really didn't have much choice, with John being the actual skipper of the ship which gives him the authority to marry them."

"Hey," Miranda snapped, as she leaned down and looked under his frowning face, "What's wrong? You okay?"

Not shifting his look in her direction, he just said, "Oh yeah. Couldn't be better."

Miranda jerked her head back up while looking down at him, thinking of just how much effort she put into fixing herself up when he called her to ask if they could meet up there.

She started walking off quickly, and remarked with anger in her voice through the fast clicking of her spiked heels as she got halfway to the elevator, "Well, I certainly hope you and this woman you're head over heels for finally hit it off together when we get back home!"

Just as she popped the elevator button so hard it almost broke, Jerry shouted, "Miranda! Please stop and hear me out!"

The elevator door quickly opened, and she stepped in, spun around while folding her arms and let her back impact against the rear wall of the elevator, and shouted back, "Go cry on someone's else's shoulder!"

Just as the elevator doors began to close, Jerry yelled out just in time before they shut completely, "This woman's not back home! She's onboard this ship!"

The doors then shut completely. Jerry turned back around, leaned his elbows down on the arm rests against again, and tears began to flow down his puffy cheeks.

A few moments later, the silent bell at the elevator gave off its soft tone, indicating someone had arrived. After the doors opened back up, Jerry could hear the slow walking click of high heels gradually walking toward him. They stopped just as she now stood next to him again, but he was too embarrassed to look back up to her with all the tears now coming from his eyes like a waterfall.

Miranda softly asked, "Jerry? Would you please look at me?"

He shoved himself up off the railing, and gradually turned to her, and almost went into shock. Tears were pouring down her face as well.

"Look," he said stuttering while wiping his face off with his shirt sleeves, "I'm so sorry...."

She grabbed his arms and flung them apart, wrapped her arms around his neck and pulled her face up to his, and kissed him so passionately his love for her went off like fireworks.

He wrapped his arms around her, kissing her back just as obsessively that they both lost track of all awareness.

They backed off for a moment for air, and she said with such a grand shining smile, "How dare you," and then pulled his face back kissing him with even more enthusiasm.

When she backed away again, her face was gleaming so brightly like no one had ever seen, and he asked, "How dare I? What do you mean..."

"How dare you not tell me, Dr. Jerry Danielson! I've been in love with you from the first moment I laid eyes on you! That night we were in here and you said you hadn't been with another woman for a little over three years, my heart exploded when I did the math realizing that's about the time we met!"

"Oh, my God Miranda! I've never, ever been in love with someone like I have you! In fact, I never really knew what true love really was until the boss introduced us, that's why I didn't know what to do! I thought and thought for days just how in the world do I handle this, but that's when you started throwing all those pleasant remarks at me, and I knew it was because of my not so nice track record."

Both still crying their eyes out, she said sobbing, "Jerry? Have you ever heard of jealousy? I knew about your so called, 'track record,' your Dr. Jerry Danielson, 'Fan Club,' all the computer links from all the big periodical publishers with you on the cover saying, 'This Month's Sexiest Man Alive,' or anywhere we went for public appearances for the brass, I noticed you turned every head from every woman that so much as got even a glimpse of you!

"Not to mention even some of the men that couldn't take their eyes off you. And here I am so in love with you I would excuse myself to run to the restroom and cry my eyes out. But what made me love you even more is how you always politely shooed every one of them away."

"Did it ever cross your mind as to why," Jerry exclaimed, "Because I knew the only woman I ever really loved my whole life was there, and I was determined not to let anyone get within a certain distance to me. Why? Because I loved, admired and respected you so much, there was no possible way I'd ever let you have an image of me with anyone else."

"Moments," she said so softly to his tearing up face again, "You do have," and then pulled themselves together once again kissing into a blissful trance.

That is until they barely heard a small thump coming from somewhere across the room, immediately followed by a restraining whisper, "Ouch, Barbara, that hurt!"

"Oh well," Barbara shouted from the same dark place, "Now that you've given us away,
we might as well join the party!"

Barbara then came running out from where her and Steven were hiding, crying her eyes out as well, yelling, "Oh, Mir! I knew it! I just knew it," and jumped and almost tackled Miranda, hugging her so tight she was so happy for her.

Jerry's tall figure towered over them laughing, and said, "Spying on us, eh? Come here, my little baby sister!" With Barbara's little petite shape, Jerry actually had to pick her up so she could get in a good hug from him.

While sitting her back down, she started saying, "I've been telling her for years how you really felt, and she absolutely refused to believe it, saying there's no way the most gorgeous man in the world would ever go for me!"

"You knew," Jerry laughed, "All this time Barb?"

"Not actually," she replied, "She told me, what was it Mir, three days after you and Jerry were introduced?"

Another voice from the opposite side of the darkened room startled everyone as John's voice leaped out of nowhere that shouted, "And he thought just the opposite Miranda! Now, what am I going to do with you two?"

Miranda shouted back, "You knew too?"

"From the time just after we first left earth," John answered, "Remember when you two came storming in here because Steven thought I was kicking Jerry to death?"

"Kind of hard to forget," Miranda said, "You Barb?"

"Of course. When they were really talking about women, and they cleaned themselves up quickly from laughing so hard just before we got here, pretending like nothing was wrong."

Miranda looked up at Jerry, and asked with a touch of humor, "That's why you two were laughing hysterically? Because you told the boss how you really felt about me?"

Jerry quickly replied, "No! No, of course not. It was about," He then glanced at John, who was wearing a grim smile with his arms fully folded, that Jerry read as, 'don't you dare', and continued, "Something that happened to someone else. That's the truth."

Barbara then said, "It's about something that happened to the boss man here, if I'm reading my signals right!"

"You're signals serve you well," John said, "Okay, I knew you two were meeting here tonight because Jerry asked you to, just to see if he was finally going to tell her. Well, I'm so very proud of you Jerry for finally telling her."

Miranda turned from Jerry's side, and as she wrapped her arms around him, she pressed the side of her face into his chest, and said as she began to cry again, "Not as proud as I am. And after all those horrible remarks I threw at you all these years," and then squeezed him as hard as she could.

John then said as he turned and began walking toward the elevator, "I'm calling it a night guys, so let me get on the elevator, and once the doors are fully closed you may tell them what happened in my case Jerry. Just please wait until the doors are all the way closed!"

From all the way down in his quarters, just as John began to climb into bed, even through the ship's bulkheads he could hear a faint echo of a tremendous outburst of such hysterical laughter. It went on for at least fifteen to twenty minutes. He quietly laughed along with them, until he gradually slipped off to sleep.

Another week later, John had the special honor of bringing not just one, but two couples together in the marital bonds of matrimony.

CHAPTER 16

MESSAGE IN A BOTTLE

Five weeks after the two weddings, lying in bed that night, Jerry said, "Well, tomorrow about noon we will pass the halfway mark. Six more weeks until we get back to our universe, our universe anyway, and three more weeks from there we'll be home. But just for my emotional security, before we get to the Hubble station I want to swing around to Uranus and see the moon Titania still intact, and that Oberon isn't!"

Miranda laughed for a moment, and said, "Ditto on that!"

She then asked, "So when we do get back home, are you absolutely sure you wouldn't mind living in my house baby?"

"Anywhere we live sweetheart," Jerry said, "Just so long as we're together, is fine with me. Besides, we'll probably be out exploring the universe most of the time."

Miranda snapped back laughing, "Just so long as it's our universe!"

"Ditto on that, Mrs. Danielson!"

"God," Miranda exclaimed, "I love the sound of that, Mr. Danielson!"

She then rolled over on her side, propped her head up with her hand resting on her elbow to face him while he was looking up at the ceiling in total bliss, and remarked, "I still can't believe the other me had a maid, and a beautiful little Pomeranian named Mojo. You know, now that I think about it, I always wanted a cute little Pomeranian. Just didn't figure I'd never have any time to spend with him. Maybe that's why I, or she rather, has a maid. Just to take care of him."

"Then when we get back, let's get one," Jerry said, "We can even name him Mojo, if you want."

"Sounds great honey," she replied, "You are much too sweet to me, but we have plenty of time to think about it. Hey, you want to go up to the bridge, and see what Barb and Steven are doing?"

"Sure baby," Jerry replied, "Let's go see if Steve's smile is still plastered across his face since after the first night of their honeymoon!"

Miranda said laughing, "You mean like ours?"

"No one's baby," Jerry said laughing back, "Will ever be as great as mine still is!"

"Well, Mr. Danielson, I'll just have to make sure it always stays that way! C'mon, maybe we'll catch them smooching!"

Walking into the bridge, John was sitting there with them talking.

"Well," John shouted, "Look who came out for some fresh air!"

"And just what are you three up to," Miranda asked, "Talking about how impressive the view is?"

The five of them then all looked out of the main viewing window together where nothing but the cold empty void of darkness encased them, that seemed like they were all just drifting through nothing but an endless eternity.

Jerry then looked back at Miranda with an unusually distracted look on his face, glancing out toward the window, and then back to her. Each time he did the distracted face looked more confused by every moment.

Miranda then asked, "Jerry? Are you okay?"

"Baby," he finally answered, "I told you a little while ago that tomorrow around noon we'll be passing the halfway point, right?"

"Yes," she replied, "So why the weird look on your face?"

"Holy cow," Jerry said with a frightening sound, and then walked over closer to the massive window and stood frozen stiff with his glare locked out toward the black emptiness.

"Baby," Miranda sharply remarked to him, "What is it?"

Still unmoving, John then followed up with Miranda's question, "Jerry? Are you alright?"

He then spun back around to the four of them, and said firmly, "No, I'm not. Guys, there's something that just hit me that I can't believe never dawned on probably any of us since we discovered we were not in our own universe. Now listen to me very carefully."

All eyes were now locked on him in deep concern as Jerry began pacing slowly back and forth in front of the viewing window.

"Okay," he finally said, "Has it occurred to any of you since that moment of waking realization that we were not home, and we all instinctively focused all our thoughts in one direction that there was something we had and must do, and that the eerie fear of actually where we actually were compelled us all to keep our minds in tunnel vision to do only one thing until we got back to our real home?"

"Yeah," John replied, "Get the hell out of there as soon as we could before they realized we weren't…us? Whoa, Jerry! Now it's dawning on me too! I see where you're going with this, and can't believe I never thought of it either!"

Barbara then demanded, "Alright! You're both starting to scare me here! Would one of you please get to the point and let the rest of us in on it?"

Steven's voice then sharply intervened, and instantly said as he rose from his chair, "I know exactly what they're talking about, and can't believe I didn't think of it either. Wow, this is one heck of an awesome concept!"

Miranda then shouted, "Then would one you please share it with Barb and me? Now I'm scared!"

"Sweetheart," Jerry shouted while quickly walking over to her until he was looking directly into her frightened face, "The other five of us! Where are they? Did they actually make it to our universe as we did theirs, and if so, what the hell happened when they did?"

"He's right Mir," Barbara said as she too got up and began pacing, "Oh man, what a bizarre thought!"

Jerry swung around the pilot's station, and quickly leaped into his pilot's chair. His arms began moving so fast again across the instruments it was almost like he was on instant scramble.

"Barb," he shot out fast, "I have the exact moment recorded here of back when we were headed in the opposite direction when you were up here by yourself, and that signal that shot passed us so fast our censors were unable to pick up any info at all on it. So you immediately reversed the long range scanners to at least try and get some kind of reading on it, right?"

"Right," she agreed as she was now standing right next to him, "but it was moving so incredibly fast, by the time I boosted the long range scanners up to their maximum distance, it was too far out of range to detect anything."

"Good girl, little sister," Jerry remarked, "You instincts always rock so well!"

The other three were all standing behind Jerry's seated position with Barbara now sitting in the co-pilots chair.

She swiftly said back, "But I didn't get anything on what it was."

Jerry smiled as he countered her remark, "Oh, my sweet little sister, yes you did! Maybe not to get a scan of what it was, but at least enough to get a course direction and velocity reading on it, which I'm sure that maybe not any of the auto censors did, but the long range scanners I'm positive had to at least get something on that."

Everyone stood still in complete captivity watching Jerry's gifted moving hands sail through all the dials and switches the control panel could provide.

What seemed like a few long moments later, he happily shouted, "Bingo! Way to go Barb!"

She shouted back, "What? Did I actually get something on at least that?"

"You most certainly did my dependable ace co-pilot! God you're good! That quick instinctive thinking of yours just gave us what we needed to know. The auto censors couldn't move fast enough to get anything, but when you instantly reversed the long range scanners, just in the last tenth of a second our long range eyeballs caught a tiny piece of its direction and speed in that final split second just before it instantly vanished out of range!"

John leaned down between them, and then asked, "Which are?"

Jerry smiled as he pointed to John both readings now up on the control panel, "Seven thousand six hundred sixty five light years per second, our identical speed at exactly that same time, and traveling exactly by the numbers in the direction of our universe!"

"Then there's no doubt sir," Steven confirmed, "It had to be the other five of us. That's almost too unbelievable to accept."

Miranda then remarked, "Why does that seem so impossible Steven? I mean think about it guys. The Barry Harper, General Vaughan and almost everyone we came into contact there were almost identical to the ones we know. I mean even the ship they built for us was more than likely identical, even the name, because someone there would have discovered any differences about it when we got there.

"The powers that be there, along with ourselves, all thought we had somehow returned. So they throw this party for us, and no one there thought of us were any different, nor did we about them. Right?

"What I'm getting at is everything there was practically the same, and in relative concepts only some minor exceptions were different. Like this little Pomeranian and that maid the other Miranda had. I've always wanted a Pomeranian, but never got around to getting one. And as far as the maid, there's no way I would have him put in an animal facility when I had to leave for weeks and months at a time, so I would have hired a live-in maid mostly to take care of him.

"Your mother Steven died three years ago, but the other Steven's mother only passed away only a few weeks ago. The only big difference, is that their General Vaughan's first set of idiots blew away the Uranus moon Titania, whereas ours wasted Oberon."

"I see your point Miranda," John said as he sat back down behind Jerry, "So odds are our counterparts picked up these same differences, and centered all their concentration on returning back to the safety of their home like we did, and chances are they're on their way back to their universe like we're doing."

"Hopefully," Jerry added, "But here's what I'm thinking. Tomorrow just before noon we will be half of the way back home, which was seven minutes prior to what we have calculated to be the halfway crossover point.

"I doubt they discovered what they had to do precisely at the exact instant we did, but if they're characters, personalities and ways of thinking are almost identical to ours, they had to come across these minor differences close to the amount of time it took us to realize it too.

Jerry then walked over to main viewing window staring into it like he was looking for something, turned back around and continued, "Anyway, my instincts are betting good odds that we'll pass them once again at some point before we get back, and if my guess is right, sometime in the next few days. Even tomorrow maybe."

"Makes sense so far Jerry," John said, "Because once they do my emotional security will rest much easier knowing when we finally get home, they won't be there.

But every day for the rest of our return journey after tomorrow, if we don't cross paths with them, I'll start having panic attacks that will get worse by each passing day."

"Oh my God," Barbara joined in, "We better pass them, and the sooner the better, because if we don't it could mean our people discovered they weren't us and have them all locked away somewhere. I have no intention of meeting up with my counterpart because for some reason they didn't act in time and couldn't get away before our powers that be had them all put away somewhere. Meeting up with the other Barbara would freak me out to the point of hysteria."

"It would mentally change me," Miranda agreed, "I'd never be the same again if I met up with mine."

Steven then said, "I would have to be permanently institutionalized, incapable of rational thought."

The bridge went dead silent, with everyone staring at Steven.

"Alright, alright everybody," Jerry said laughing, "I didn't bring this point up to get you guys start thinking of a plot for some horror movie. I'm sure they all feel the same about having to meet up with us, and I'm almost positive they acted just as we did."

"You brought it up for some reason Jerry," John remarked, "So let's have it."

Jerry sighed for a second, and then said, "I've always trusted my gut instincts, with the exception of Miranda. The overpowering love I've always had for her blew all my thoughts where my instincts generate from all to hell. Other than that, they've never been wrong, and I know we're going to be passing them real soon now.

"So here's the reason I brought this up. I say as they pass us, we tell them hello in some kind of way."

The whole bridge went dead silent again. Everyone then started looking at each other, wondering if the exposure to deep space had finally gotten to Jerry.

Barbara then asked, "Well then what's the plan, big brother? Open one of the hatches, and stick your arm out and wave at them?"

"Wait a minute Barbara," John said, "He's got something here. Maybe not a real life hello, but some type of acknowledgment to them? Say an electronic flashing pulse of some type, or something like that to let them know we got away as well?"

"Bingo, boss," Jerry agreed, "You read my mind exactly. Just as a nice gesture to them. After all, in their universe, they're the top heroes there like we are in ours. I just think it would have some type of historical significant value some day in the far distant future, and it would be pretty cool of us to be known as the ones that thought of it."

"But Jerry," Steven said, "What if they take it to be a signal that they want to meet up with us, and turn around and follow us?"

"Think about it Steven," Miranda said, "They're us in a big sense. Would we turn around and start following them? I think it's an awesome idea baby. I mean the way both of our technologies have progressed so rapidly on a geometric scale, our civilization will definitely meet up again with theirs someday. And I really don't believe it will be that far in the distant future either."

"I like it," Barbara added with even more enthusiasm, "After all, our names, especially my sweetie's and mine, will come to be known for uncounted generations to come. So will our counterpart names."

"How about it Steven," John asked, "You and your soul mate are the computer experts here. Can you guys rig up something, say in the long range scanners to send some type of signal to follow our course ahead as far ahead as you guys can possibly make it stretch?"

Steven began to slowly shake his head yes, and then said, "Yes, it's possible. Very possible. Barbara? What do you think?"

"Of course," she replied, "And I know just how we can do it baby. Remember what Mr. Harper said at our first celebration dinner? He said, "Our magnificent discovery can do much, much more than propel spacecrafts, my friends. The kinetic energy given off by the imploding tachyons can be so easily channeled to power just about anything."

Steven just rolled his eyes up, and said, "That's it! We'll just set up any type of signal you wish sir, and channel it to ride on a tachyon beam emission. Let's see something here real quick."

He reached into both of his pants pockets, and pulled out his two hand operated little computers. Only using his two thumbs, they flew across the keypads so fast it looked as if everyone was watching him on a monitor screen moving on fast forward.

Steven then stopped, spun both hand computers around like a gunslinger as he shoved them back in his pockets.

"Okay," Steven said, "The beam will begin to fade at almost seventeen thousand light years ahead. At our present velocity, that will give them a little over two seconds when they pick it up. Good enough sir?"

John, along with the rest of them stared at him with total fascination, when John then asked, "Based on your calculations, if we shrink the message down as far as possible, how long of a message in that small amount of time do you think we can send?"

"In lower case digital format," Steven thought for a brief moment, "I'd say at least one hundred words minimum. Maybe five, even up to ten more."

"Okay, you and Barbara go do what you have to do so we can start transmitting it as soon as possible just in case they got an earlier start than we did, and the three of us will work up something to say to them. Good work, Steven."

"Thank you sir," he replied, and then looked over to his wife, and asked, "Ready honey?"

Barbara was in a state of total bliss looking at him, and sighed while saying, "What a man. Just lead the way my king."

They got up to leave the bridge for the computer room, when Miranda said, "Wait, wait a second Steven. How do you do that fancy spin with those two little computers to make them land back in your pockets like that?"

Barbara answered, "Oh, Mir, you wouldn't believe how talented he is with those two amazing hands."

They then turned to leave, when Miranda shouted, "I know another man that I just married that's just as creative with his two magical hands!"

A surge of wooing flooded throughout the bridge, with John shouting, "Alright, alright, let's keep it professional my sex crazy crew."

Miranda stood up and hugged Jerry while pressing the side of her face into his chest, and told John, "Crazy isn't the word John."

The wooing got even louder then, and John just rolled his eyes up, and stood up while pointing at Barbara and Steven, and then yelled, "Okay, you and you. Go do your genius stuff with the computers."

Turning to Miranda and Jerry, John said, "You and you, work up something to tell the other singing lovebirds. I'll be back shortly. I have to go and get rid of something now."

He walked passed Barbara and Steven with a look of frustration, and as he opened the main bridge door, Barbara asked, "What do you need to get rid of Dr. Stanley?"

John stopped with the door halfway open, looked back to the four of them, and shouted with a tone of humor, "A horrifying flashback image you love sick perverts just brought back to me!"

After slamming the door, John got a few steps away and heard an erupting explosion of hysterical laughter coming from the bridge, knowing all too well the four of them would get a real kick out of that remark.

He then smiled as he knew that image of his wife with another women was long gone, and decided the biggest priority that would be number one on his list when he got back home was to find another woman to fall in love with and marry, so that he too could finally have what he now considered his four family members had to share.

Each other.

CHAPTER 17

THERE'S NO PLACE LIKE HOME

Minutes away from the Hubble VII Station, the ***Archimedes*** took a small detour as she swung around Neptune on a new direction heading for Uranus.

"I'll explain everything once we land back at the platform Barry," John responded to his repeated requests, "There's just something we want to see first, and then we'll be right there."

Miranda walked into the bridge up to where Barbara and Steven were sitting right behind John in the pilot's seat.

She looked around for Jerry, but was startled when she heard Barry Harper's voice angrily blast out, "I can't believe this John! You guys leave for four and a half months, and then suddenly vanish the next day, and now you show back up exactly four and a half months later again? Don't expect a hug and a kiss from General Vaughan when you see him!"

"Oh, I have a feeling both of you will once you hear the full story, old buddy! We'll be there in a little while, but first we want to be absolutely sure we're home this time," John replied .

Barry shot right back, "By doing what John?"

"By taking a quick look at one of Uranus' mighty moons, my good man. See you shortly!"

He then looked over to Miranda, and then asked, "Mind going up to the observation deck and telling your new hubby it's time for him to take over and bring us back to the Hubble landing platform?"

"Sure," and walked over toward the elevator.

Walking out she saw Jerry leaning on the padded support rail, gazing out of the huge observation deck dome.

Jerry leisurely turned in her direction as she walked up and kissed him on the cheek, and said, "The boss asked me to let you know he would like for you to come down and take over your mighty pilot's chair, to swing us back around once we verify that moon's still intact and bring the ship to the Hubble station."

He still had that skeptical, suspicious grin painted across his face, the same one he had for weeks now whenever he was near his new wife.

He just smiled again at her, and replied, "Yeah baby. I'll be down in a minute."

She slowly wrapped her arms around him, and hugged him hard while she asked, "Is that naughty boy grin ever going to wear off sweetheart?"

Hugging her back, he replied, "You're never going to admit it, are you?"

She swung her beautiful glazing eyes up to his, and said, "C'mon, baby. You really need to let that go."

Miranda then rose up to kiss him.

Lowering herself back down, he smiled and said, "Didn't think so."

"I love you," she said, "Very, very much Jerry. What difference would that make now?"

"That's what I keep telling you," Jerry answered, "It wouldn't change the way I feel about you, one way or another baby. I just want to know the truth, that's all."

"Good," Miranda said, "Glad that's finally over."

"I didn't say it was over baby. I just said the truth wouldn't make it matter."

"You said," Miranda countered, "One way or the other."

"Because if she did," Jerry said firmly, "I think that would have been pretty cool of her to do that for us. Otherwise, we would maybe have never known."

"Maybe," she replied, "But whatever currents control the outcome of the Cosmos, it did in fact happen."

"But what if the currents of the Cosmos blew in the other direction," Jerry said, "I mean God knows how many times I worked up the courage to tell you, and backed away each time. I was so scared from all the ugly comments you threw at me for three years, I kept thinking that there's no way she'll go for me."

"Baby, Barbara has been telling me for over three years now how you felt about me, and I told her three days after we were introduced how I felt about you. I just took a shot at it, figuring what did I have to lose? By now he would have said something!"

Jerry then reached out and started tickling her, saying, "C'mon, let's have the real truth now or I'll tickle you into convulsions!"

Miranda was wiggling and screaming hysterically, yelling, "Stop! Stop! Help!" But Jerry was much too strong for her to get away.

The elevator doors opened and Barbara stepped out then Steven. Steven saw what was going on, and immediately did a U turn back toward the elevator compartment, but Barbara reached back instantly and pulled him back out.

Jerry was holding her upright as she was catching her breath back, when Barbara walked up laughing and shouted, "What are doing to her, my big bully brother?"

Miranda said through her short breaths, "He's still at it Barb! Trying to get me to admit something that never happened!"

Barbara shouted, "C'mon Jerry! Why concern yourself with that situation still? Do you think by now you would have ever said anything if she hadn't?"

Miranda then said, "Why baby, would I lie about something like that anyway?"

"Ego," Jerry said, "I know she'll never admit it if it were true."

"Just pray Jerry," Barbara said, "That it was God that arranged for it to happen, and that He was only helping you."

"I know," Jerry admitted, "At least it did happen."

Barbara added, "I mean it was God that brought Steven and I together…Steven? Where'd he go?"

"Over here," his voice squeaked out from the back wall, "Is everything okay now?"

Miranda said over to him, "Yes, it's safe now Steven. He was just tickling me. That's all!"

Barbara said in a low whisper, "He's got a phobia about being tickled. I tried it once on him, and he threw up. Anyway big brother, the boss needs you back at the helm. When Miranda didn't come back with you, he sent us as back up."

140

"Alright," Jerry sharply said, "I'm just going to say this once more, and it's over. I promise baby."

Jerry then took a deep breath, and said, "Just as we passed the halfway mark in the black void coming back, we get this message from our counterparts two seconds before they zipped by us. Miranda was on watch, and wakes everybody up to get to the bridge pronto.

"I walk in, and Barbara and Steven are examining this message from them, that you Miranda was the first to receive. I walk up and ask what the big deal is, and you baby look at me smiling like a kid in a candy store, and you walk by me grinning from ear to ear saying you don't know, ask them, and you take off out of the bridge saying you had something to take care of, that Barbara's going to take over the rest of your watch.

"So I turn to Barbara and Steven just as John comes walking in, and you guys read the message to John and I. It stated, "Hello, we are your counterparts from the other universe Hope you guys caught the minor differences in time to get back home as we did. Go in peace for all mankind. One more thing…" and that's where it stops.

"Steven clearly tells us that somehow the rest of the message, which consisted of three more sentences, was deleted upon our instruments receiving it. So Barbara checks the backup recording log, and it had been erased there too!"

Jerry steps closer to Miranda, gives her that suspicious glare, and Jerry said, "Since I was worn out, I said I'll celebrate when I woke up. Exactly forty five minutes later, you knock on my cabin door, and when I open it you just barge right in with your robe on with an ear to ear smile, and you even order me to lock my door.

"Once I do, I turn back around and the robe's on the floor, and you're standing in my cabin all decked out like Playmate of the century, and ask me if I was really in love with you. That, didn't require even math to figure out sweetheart. But I'm glad you did."

Jerry then turned and began walking, saying, "Let me go resume my post before the boss comes hunting for me."

Halfway to the elevator, Jerry gestures his arm up toward the vast star field that glittered through the glass dome, and shouts, "Thank you Miranda, wherever you are out there."

Miranda just stood there with Barbara and Steven, shaking her head like he was crazy.

Miranda just sighed, and said to them, "He'll never truly accept that the other Miranda didn't say that."

Barbara just reached out and hugged her, and said in a low whisper, "We're tired, and going to lie down before we land back at the Hubble Station. By the way, you rock big sister."

"Sweet dreams, little sis," Miranda said to her and Steven while turning back around to lean on the cushioned support rail.

She then noticed the *Archimedes* now slowly beginning to adjust its course to head them back to the landing platform on the Hubble VII station, and avoid getting any closer to beautiful little shepherd moon of Uranus.

The moon known as Oberon.

Once she noticed the elevator door close completely with Barbara and Steven, she turned back forward, raised her head up out at the heavens, and then smiled with an explosion of joy.

She then reached into her pants pocket, and pulled out and opened up a small sheet of paper, with a text in lower case digital format, that she printed out from the ship's long range receiving monitor that she had removed from the monitor's printer, and erased it from the receiver's backup recording log, with a few sentences imprinted on it.

It stated, "One more thing. Miranda, if he hasn't told you yet please forgive him because he thinks you hate him. Jerry Danielson is head over heels in love with you and has been for a little over three years now from the day he met you and the most fantastic loving man in the universe and I am so proud to be his wife. Godspeed and good luck with him.

    Miranda Young Danielson."

She then quickly folded it back up, and shoved it back into her pants pocket.

Miranda then looked into the star light glowing in from the glass dome above, and said in almost silent whisper, "Thank you Miranda, wherever you are out there!"

****

www.ingramcontent.com/pod-product-compliance
Lightning Source LLC
Chambersburg PA
CBHW061511180526
45171CB00001B/129